ECOLOGIES OF THE EARLY GARDEN CITY

Essays on Structure, Agency, and Greenspace

Graham Livesey

ECOLOGIES OF THE EARLY GARDEN CITY

Essays on Structure, Agency, and Greenspace

Graham Livesey

COMMON GROUND RESEARCH NETWORKS 2019

First published in 2019
as part of the Constructed Environment Book Imprint
doi: 10.18848/978-1-86335-129-4/CGP (Full Book)

BISAC Codes: NAT045000, ARC010000, GAR014000

Common Ground Research Networks
2001 South First Street, Suite 202
University of Illinois Research Park
Champaign, IL
61820

Copyright © Graham Livesey 2019

All rights reserved. Apart from fair dealing for the purposes of study, research, criticism or review as permitted under the applicable copyright legislation, no part of this book may be reproduced by any process without written permission from the publisher.

Library of Congress Cataloging-in-Publication Data

Names: Livesey, Graham, 1960- author.
Title: Ecologies of the early garden city : essays on structure, agency, and
    greenspace / Graham Livesey.
Description: Champaign : Common Ground Research Networks, [2018] | Includes
    bibliographical references.
Identifiers: LCCN 2018047246 (print) | LCCN 2019005884 (ebook) | ISBN
    9781863351294 (pdf) | ISBN 9781863351270 (hardback : alk. paper) | ISBN
    9781863351287 (pbk. : alk. paper)
Subjects: LCSH: Garden ecology. | Landscape ecology. | Urban
    gardening--Social aspects. | Open spaces.
Classification: LCC QH541.5.G37 (ebook) | LCC QH541.5.G37 L587 2018 (print) |
    DDC 577.5/54--dc23
LC record available at https://lccn.loc.gov/2018047246

Cover Photo Credit: Garden City Collection

Table of Contents

Acknowledgements xi

Introduction 1
    Assemblage Theory 2
    Organization of the Book 8

Part 1: Landscape Ecology

Chapter 1 11
Flows and Turbulence
    Defining Flow 11
    Human Constructed Flow Systems 14
    Turbulence and Habitats 18
    Conclusion: Productive Flows 12

Chapter 2 25
Patches and Patterns
    Patch-Corridor-Matrix Spatial Model 25
    Urban Patchwork Dynamics 27
    Evolving Patchwork Patterns 30
    Assemblages and Territories 32
    Conclusion: Activating Ecologies 34

Chapter 3 37
Boundaries and Corridors
    The Functions of Boundary Systems 37
    Boundary Dynamics and Ecotones 40
    Corridors and Networks 44
    Conclusion: Assemblages and Boundaries 46

Chapter 4 49
Agents and Agencies
    Assemblages and Agency 49
    Individuals versus Organizations 52
    Organizations and Boundaries 54
    Conclusion: Towards Boundaryless Organizations 56

Part 2: Early Garden City

Chapter 5 61
Garden City Theory
    Influences on the Garden City 61
    Ebenezer Howard and *To-Morrow: a Peaceful Path to Real Reform* 68
    Conclusion: A Rural-Urban Vision 75

Chapter 6 77
Letchworth Garden City, 1903-1913
    The Early Development of Letchworth 77
    C.B. Purdom and a Critique of the First Decade 81
    Conclusion: Garden City Innovations 85

Chapter 7 89
Gardeners and Gardens
    The Qualities of the Gardener and Gardening 90
    The Garden as an Assemblage 94
    The Evolution of the Park 96
    Conclusion: Beyond Maintenance 98

Chapter 8 101
Farmers and Greenbelts
    Farms and Farming 101
    The Functions of Greenbelts 105
    Benefits and Problems 108
    Conclusion: Active Ecologies 112

Chapter 9 113
Letchworth Garden City Innovations
    The Management of Flows 114
    The Functional Organization of Land 118
    Boundaries and Corridors 123
    Agents and Agencies 128
    The Gardener as a Community Figure 135
    Greenbelts and the Role of the Farmer 141
    Conclusion: Letchworth After 1913 146

Chapter 10 151
Conclusion

Bibliography 159

Acknowledgements

I owe debts of gratitude to the following people and institutions:

First, to Professor Arie Graafland and the Faculty of Architecture at the Delft University of Technology (TUDelft).

Second, I would like to thank those who carefully read drafts of the material at various stages of the process, particularly Dr. David Monteyne at the University of Calgary, Dr. Lara Schrijver at TUDelft, and anonymous reviewers.

Third, I would like to acknowledge that portions of various chapters have been previously published in earlier versions, and that input and encouragement was provided by Dr. Adrian Parr (University of Cincinnati), Jo Odgers (Cardiff University), Dr. Diana Masny (University of Ottawa), Daniel Pettus (Claremont Graduate University), Dr. Mark Dorrian (University of Edinburgh), the organizers of the "Agency" conference held at Sheffield University in November 2008, and the editors of the *Berkeley Planning Journal*.

Fourth, the staff of the First Garden City Heritage Museum in Letchworth.

Fifth, colleagues, staff, and students in the Faculty of Environmental Design at the University of Calgary.

Sixth, the staff at Common Ground Research Networks.

Finally, and above all, to my wife Allison, and our children and families.

Introduction

This study examines various urban ecology topics derived from the field of landscape ecology, this is supplemented by material from the complementary concept of assemblage theory. In a somewhat speculative manner the theory will then be applied to an interpretation of the first decade (1903-1913) of the pioneering Garden City at Letchworth in Hertfordshire outside of London. The Garden City, a deliberate effort to unite town and country, emerged as a direct response to what were perceived to be the evils of the large industrial cities of the nineteenth century and attempted to reunite country and town through the garden, particularly the role of the private garden and the act of gardening. The early Garden City movement, as represented by the writings of Ebenezer Howard, particularly his book *To-Morrow: A Peaceful Path to Real Reform* published in 1898, was a concentrated attempt to restructure the formal and spatial aspects of a town, and to redefine many urban practices. The inauguration of the Garden City was a significant development in urban history, particularly in the first half of the twentieth century; the movement would inspire a broad range of urban typologies and various visionary models of the city during the twentieth century.[1] The Garden City legacy also provides a foil for examining a number of questions regarding contemporary urban ecology. What are the ecological factors effecting cities in terms of urban structure? What are the agencies that effect change in cities? How do gardens and farms function, and what operational role do gardeners and farmers play in urban systems?

At Letchworth, the designers and builders of the town attempted to produce a synthetic model that addressed a broad range of urban issues from structure, spatial organization, housing design, civic education, to political, social and cultural organization. As an experiment in the design of a total community it was relatively successful, despite the many years it took to reach its target population. The agricultural belt has had an enduring and important influence on urban design and regional management since its inception, the emphasis placed on gardening at Letchworth has been largely lost. The advent of many new green space typologies during the twentieth century would do much to reterritorialize the relationship of a city to its region, and also would aid in the significant impact of green space networks on urban structures and organizations. The relatively low density of housing at Letchworth (12 units per acre) is formally very similar to the suburban communities developed globally since the Second World War, however, suburbia is distinctly different from the Garden City model as the Garden City was intended to be self-supporting, with a full range of employment opportunities and amenities. The Garden City proposed a radical reorganization of urban structures and community organizations, with a concerted effort to challenge capitalism and the private

---

[1] See R. Freestone, "Greenbelts in City and Regional Planning," in K.C. Parsons and D. Schuyler, eds., *From Garden City to Green City: The Legacy of Ebenezer Howard* (Baltimore: Johns Hopkins University Press, 2002).

ownership of land. As a satellite to a larger city the Garden City was, in part, designed to provide a familiar environment to those migrating to the city from the country. In practice it was attractive to pioneering professionals looking for alternative ways of living, and to factory workers seeking a better place to live.

C.B. Purdom comprehensively documented the early development of Letchworth in his book *The Garden City: A Study in the Development of a Modern Town*, published in 1913.[2] He provides detailed information on Howard's theories, and of the history of the development of Letchworth. The center of the community is occupied by a small-town square, private gardens and common areas define the low-density residential neighborhoods, and the town is encircled by an agricultural belt (or greenbelt). All of this was intended to give the inhabitant the sense of living in the country, and in an environment that is healthy and clean. The major innovations were the agricultural belt, which would transform over time into a wide range of new urban typologies, and the emphasis on private gardens and the act of gardening. Letchworth was conceived around the experience of being in a garden,[3] at both the intimate level of the private garden and at the communal level of the town. The use of the agricultural belt, or greenbelt, as a wide boundary separating the town from the country and providing a protected zone for agricultural and cultural amenities, helped create the sense of a town engulfed in greenery. This is augmented by other green spaces in the town itself and enhanced by the generally single-family housing fabric and the emphasis placed on private gardens.

The theoretical aspect of this study is informed primarily by landscape ecology (particularly the writings of Richard T.T. Forman), but also by Gilles Deleuze and Félix Guattari's concept of "assemblage," which will be briefly presented here. The conjunction of landscape ecology and assemblage theory provides a framework for examining a number of ideas associated with the early Garden City movement.

ASSEMBLAGE THEORY

The concept of assemblage is central to Deleuze and Guattari's text *A Thousand Plateaus: Capitalism and Schizophrenia* and is the English equivalent of the French word *agencement*, or the "process of arranging, organizing, fitting together." [4] According to Deleuze and Guattari there is both a horizontal and a vertical axis associated with assemblages. The horizontal axis deals with "machinic assemblages of bodies, actions and passions" and a "collective assemblage of enunciation, of acts and statements, of incorporeal transformations of bodies."[5] The vertical axis has both "territorial sides, or reterritorialized sides, which stabilize it, and cutting edges of

---

[2] C.B. Purdom, *The Garden City: A Study in the Development of a Modern Town* (London: J.M. Dent & Sons Ltd., 1913).
[3] Ibid., p. 113.
[4] J. Macgregor Wise, "Assemblage," in Charles J. Stivale, ed., *Gilles Deleuze: Key Concepts* (Montreal & Kingston; McGill-Queen's University Press, 2005), p. 77.
[5] G. Deleuze and F. Guattari, *A Thousand Plateaus: Capitalism and Schizophrenia* (Minneapolis: University of Minnesota Press, 1987), p. 88.

deterritorialization, which carry it away."[6] Further, they write "an assemblage, in its multiplicity, necessarily acts on semiotic flows, material flows, and social flows simultaneously."[7] They provide a relatively precise definition when they write:

> We will call an assemblage every constellation of singularities and traits deducted from the flow—selected, organized, stratified—in such a way as to converge (consistency) artificially and naturally; an assemblage, in this sense, is a veritable invention.[8]

Assemblages are produced from larger contexts, by bringing together sets of forces, bodies, languages, and territories; assemblages are "diverse things brought together in particular relations."[9] Beyond things, assemblages include qualities, actions, emotions, languages, and territorialities. Assemblages include machinic and semiotic systems, or systems of bodies and signs; moving between technology and language, they produce new ways of speaking and meaning.[10] Assemblages create, and operate with, territories that are in constant processes of deterritorializing and reterritorializing. The territorial aspect of assemblages has particular importance for human strategies of engagement with the land, whether nomadic or settled. An assemblage brings together a set of structures, expressions, and territories in an arrangement. Ultimately, assemblages are functional, in that they create new ways of speaking or behaving, new organizations or expressions, new spaces or structures. This reinforces the notion advanced by critical theorist J. Macgregor Wise that "we do not know what an assemblage is until we can find out what it can do. Assemblages select elements from the milieus (the surroundings, the context, the mediums in which the assemblages work) and bring them together in a particular way."[11] Functional and territorial transformation is characteristic of assemblages.

Therefore, we can understand assemblages as machinic, enunciative, and territorial. Assemblages are compositions of desire and speech, they operate through "constellations" of objects, bodies, events, languages, and territories. Assemblages operate for varying periods of time to create new functions. An assemblage can be defined by a "diagram" that codifies its operations and defines the relationships between a particular set of forces that are brought into play; the assemblage diagram is, according to Deleuze, the "map of destiny."[12] Effectively, the diagram is the arrangement of forces by which an assemblage operates, it is a map of the function of an assemblage. In his book on Michel Foucault, Deleuze provides a brief description of the diagram, which he equates with his own concept of the "abstract machine," he writes:

---

[6] Ibid., p. 88.
[7] Ibid., p. 22-23.
[8] Ibid., p. 406.
[9] Wise, "Assemblage," p. 78.
[10] Ibid., p. 80.
[11] Ibid., p. 78.
[12] Gilles Deleuze, *Foucault* (Minneapolis: University of Minnesota Press, 1988), p. 36.

every diagram is intersocial and constantly evolving. It never functions in order to represent a persisting world but produces a new reality, a new model of truth. It is neither the subject of history, nor does it survey history. It makes history by unmaking preceding realities and significations, constituting hundreds of points of emergence or creativity, unexpected conjunctions or improbable continuums. It doubles history with a sense of continual evolution.[13]

As Deleuze notes, "every society has its diagram(s),"[14] or social and political systems that function for particular ends, these can be enduring or temporary, State organized, or community based. Ideally, an assemblage is innovative and productive, as the diagram structures "the space of possibilities associated with the assemblage."[15] The diagram describes the forces that have created an assemblage, and the function, or active aspect, of the arrangement. Tom Conley suggests that the diagram provides the agent or agency of an assemblage, it is the arrangement that results in action or production.[16] Conley notes that the diagram "bears resemblance to a digital operation comprised of given codes, turns into a modulating agent, thanks to the work of the unconscious in creative speculation and action, that creates new and multiple forms and colors."[17] The diagram is the arrangement of elements in the assemblage, the relative interaction that leads to a production.

According to Deleuze and Guattari, assemblages operate through desire as abstract machines. An assemblage transpires as a set of forces coalesce, these can be triggered or abetted by human or non-human agents, or they can emerge from a self-organizing system. The role of agency in assemblages will be examined in further detail below, as well as the notion that an assemblage has an agency. Deleuze and Guattari place an emphasis on innovation or creativity, what they call the search for that which has not yet been defined.[18] Assemblages emerge from the arrangement of heterogeneous elements into a productive (or machinic) entity that can be diagrammed, at least temporarily, and often has the ability to provide an agencing quality. Assemblage theory focuses on objects, bodies, expressions, and territories that come together in productive and/or innovative arrangements.[19]

The horizontal axis identified by Deleuze and Guattari includes content and expression, which involve "bodies reacting to one another," and actions and statements.[20] As Deleuze and Guattari stress an assemblage, as an abstract machine, is diagrammatic, it "operates by matter, not by substance; by function, not by form."[21] This aspect of assemblages includes bodies, materiality, and culturally determined

---

[13] Ibid., p. 35.
[14] Ibid.
[15] See Manuel DeLanda, *A New Philosophy of Society* (London: Continuum, 2006), p. 30.
[16] Tom Conley, "Afterword," in Gilles Deleuze, *Francis Bacon: The Logic of Sensation* (Minneapolis: University of Minnesota Press, 2003), p. 146.
[17] Ibid., p. 146.
[18] John Rajchman, *The Deleuze Connections* (Cambridge, Mass.: MIT Press, 2000), p. 8.
[19] See also DeLanda, *A New Philosophy of Society*.
[20] Deleuze and Guattari, *A Thousand Plateaus*, p. 88.
[21] Ibid., p. 141.

factors such as language. The content of an assemblage involves both objects (such as furniture, architecture, and other human constructs) and organisms (and the actions and passions contained within bodies), or as Deleuze and Guattari state machines with "intermingled pieces, gears, processes, and bodies."[22] The machinic qualities of assemblages involve an arrangement, and has a productive, or creative, dimension. Addressing the machinic aspect of assemblages Claire Colebrook writes: "Because a machine has no subjectivity or organizing center it is nothing more than the connections and productions it makes; it is what it does. It therefore has no home or ground; it is a constant process of deterritorialisation or becoming other than itself."[23] This introduces a vital interconnectedness between elements in an assemblage and is the content of the assemblage.

Deleuze and Guattari give primacy to bodies over objects, and enunciation over language, they also champion the social, or collective relations, over individualism and formalism. They write that expression "becomes a semiotic system, a regime of signs, and content becomes a pragmatic system, actions and passions," assemblages are also "simultaneously and inseparably" machinic assemblages and assemblages of enunciation.[24] As they state expression is independent from content.[25] For Deleuze and Guattari, expression is about the creation of new concepts, or new relations, it is the vital and innovative (or creative) aspect of their thinking, expression is "the power of life to unfold itself differently."[26] The enunciative aspect addresses the linguistic, or semiotic, aspects of assemblages, it is the expressive function, or "regime of signs."[27] This, as Deleuze and Guattari propose, operates within the machinic assemblages.[28] Further, enunciation "implies collective assemblages."[29] Therefore, the horizontal axis of assemblages contains a complex set of elements, which are essential to functionality, creativity, and productivity. The content and expressive aspects of assemblages are related to the forces of deterritorialization and reterritorialization, recognizing that assemblages are always participating in spatial systems.

The vertical axis of assemblages addresses territoriality. Conventionally territories are defined by lines, or boundaries, drawn on the earth or inscribed in maps, however, for Deleuze and Guattari the concept of the territory is primarily defined by modes of movement. A territory is not a homebase.[30] Territories are resistant to precise definition, territoriality is created through continuous processes of deterritorialization and reterritorialization, or the unmaking and making of territories.[31] A territory is a "malleable site of passage," that while it is in a constant

---

[22] Ibid., p. 88.
[23] C. Colebrook, *Gilles Deleuze* (London: Routledge, 2002), pp. 55-56.
[24] Deleuze and Guattari, *A Thousand Plateaus*, p. 504.
[25] Ibid., p. 89.
[26] C. Colebrook, "Expression," in A. Parr, ed., *The Deleuze Dictionary* (New York: Columbia University Press, 2005), p. 93.
[27] Deleuze and Guattari, *A Thousand Plateaus*, p. 88.
[28] Ibid., p. 7.
[29] Ibid., p. 80.
[30] K. Message, "Territory," in Parr, ed., *The Deleuze Dictionary*, p. 275.
[31] Wise, "Assemblage," p. 79.

process of change, maintains an "internal organization."[32] In defining a territory, Deleuze and Guattari state that functionality is a product of a territory, rather than the more conventional inverse,[33] they write that a territory effects "a reorganization of functions and a regrouping of forces."[34] The processes of deterritorialization are vital for reterritorializing, and the notion of immanent change. This is a non-dualistic and creative operation, that allows an assemblage to continually recontextualize its relations to territories whether it moves or not.

A key aspect of territoriality, for Deleuze and Guattari, is the "line of flight," they that territorialities "are shot through with lines of flight testifying to the presence within them of movements of deterritorialization and reterritorialization."[35] Territorialities both contain and are subject to nomadic lines of flight that create deterritorialization, territories are deterritorialized and reterritorialized by lines of flight. Here, the nomadic concept of smooth space is proposed as a more creative (or productive) means of territorializing, than the rigid structures of striated space.[36] A striated space attempts to fix territories, and limit deterritorialization and reterritorialization, through the subdivision of land, infrastructure systems, bureaucracies, social and political organizations. In other words, striated space tries to provide a delimited context for action, language, and affect. The nomadic, or smooth space, concept of territoriality does not seek this notion of context or territory, territories are not the ground or context for action, they are another aspect of the productive or creative relationship that bodies can have with land. Striated space, particularly in cities, can be deterritorialized and reterritorialized by lines of flight. For Deleuze and Guattari, life is a production of lines of flight, where "mutations and differences produce not just the progression of history but disruptions, breaks, new beginnings, and 'monstrous' births. This is also the event: not another moment within time, but something that allows time to take off on a new path."[37] The line of flight is defined by nomads "who escape all territorialisation and sow deterritorialisation everywhere they go."[38] Deleuze and Guattari emphasize the vitality of nomadicism, and hence the continual potential for territories (or assemblages) to deterritorialize. However, they also stress that all territories can oscillate between smooth and striated, that they contain a mixture of the two.[39] This underscores the notion that an assemblage is the totality of forces in play, and that it has emergent properties that can invert an established condition. The transformative aspects of an assemblage's diagram can be triggered by any number of external and internal factors.

Cities belong to a multitude of extensive processes and territories, in particular to trading networks and agricultural lands. As Deleuze and Guattari state the town

---

[32] K. Message, "Territory," p. 275.
[33] Deleuze and Guattari, *A Thousand Plateaus*, p. 315.
[34] Ibid., p. 320.
[35] Ibid., p. 55.
[36] Ibid., pp. 474-500.
[37] Colebrook, *Gilles Deleuze*, p. 57.
[38] T. Lorraine, "Lines of Flight," in Parr, ed., *The Deleuze Dictionary*, p. 145.
[39] Deleuze and Guattari, *A Thousand Plateaus*, p. 474.

manages flows, many of these act against the striation of the surface of the earth.[40] If we consider a city (or town) as a very complex assemblage (of assemblages) it can be seen as arrangements of systems, elements, actions, affects, bodies, enunciations, structures, and codes. Unlike the actions of nomads, the relationship between a city and the earth it occupies is highly structured, land is over coded by the systems of urban settlement, including: regulations, systems of land subdivision, movement patterns, infrastructure, buildings in all their functional complexity, gardens, parks, spatial systems based on political structures, and social hierarchies. While cities encompass a multiplicity of assemblages, the ability to deterriorialize and reterritorialize themselves according to the potentials, intensities, and gradients in the structure, can be highly restricted. A territory, or assemblage, may or may not correspond with generally accepted (bounded) structures such as states, provinces, cities, or districts. As we know, many of these are arbitrary, and tend to be redefined by social, political, or ethnic communities. Despite the striation of space, assemblages and territories continuously operate, some of these creating smooth space conditions, others redefine historic urban structures and relationships. In more contemporary cities a wide range of nomadic forces have been at play including mechanical methods of transportation, and electronic communications systems.

Assemblage theory focuses on productivity and innovation, and on complex arrangements of objects, bodies, expressions, and territories. Like machines, assemblages come together for varying periods of time to ideally create new ways of operating. The result of a productive assemblage is a new means of expression, or a new territorial/spatial organization, a new institution, a new behavior, or a new realization. Deleuxe and Guattari's concept of assemblages applies to all structures, from the behavioral patterns of an individual, the organization of institutions, to the functioning of ecologies. The assemblage is destined to produce a 'new reality,' by making numerous, often unexpected, connections; an assemblage emerges when a function emerges. The agencing properties of assemblages depends on the ability of a given assemblage to be productive, in ecological terms this could mean that human interventions contribute, rather than detract, from environments.

We can consider that any given ecology, whether urban, rural, or natural, is at any particular moment in time an assemblage of flows, territorialities, organisms, etc. An ecology has a "diagram" that describes its structure or arrangement, its productivity, and its state. The early Garden City of Letchworth was a continuously changing multitude of assemblages, as an evolving set of machinic, linguistic, and territorial processes, and as a creative or productive entity. As suggested above the territorial operations of assemblages are complex, and can be be applied to structures such as homes, gardens, cities, and the regions they occupy. This book attempts to bring these many concepts together to study one of the most influential developments in urban history.

---

[40] Ibid., p. 432.

ORGANIZATION OF THE BOOK

Drawing from mainly from landscape ecology, with contributions from assemblage theory, I will examine aspects of the ecological structure of cities, agencies inherent within cities, the development of Letchworth during 1903-1913, and the innovations developed during the early history of the Garden City. The first part explores topics derived from landscape ecology: flows and turbulence; patches and patterns; boundaries and corridors; agents and agencies. The second part provides an outline of the theories informing the development of the garden city, an overview of the first decade at Letchworth Garden City, an introduction to the key urban figures of the gardener (and the garden) and the farmer (and the agricultural belt), and a chapter on the innovations of Letchworth Garden City. It is in Chapter 9 that the concepts introduced previously (flows and turbulence, patches and patterns, boundaries and corridors, agents and agencies, gardens and gardeners, greenbelts and farmers) are examined against Letchworth's early history. The conclusion provides an assessment of the overall project, and revisits the concept of assemblage in light of landscape ecology and the early Garden City.

Part 1: Landscape Ecology

Ecologies of the Early Garden City

CHAPTER 1

Flows and Turbulence

Landscape ecology has emerged as an independent field of study during the last several decades, drawing from disciplines such as ecology, geography, biology, zoology, forestry, and landscape architecture. The analytical techniques developed in landscape ecology initially addressed the "single elements of the network, such as patches and nodes, buffer areas, corridors and linkages; or with the dynamics of the network, such as movements, flows, migration, dispersal, fragmentation and connectivity." [1] This initial approach has since evolved into integrated and comprehensive frameworks for ecological analysis, which have typically been employed in examining the performance, or ecological effectiveness, of non-urban environments. Nevertheless, landscape ecology provides powerful theories and techniques for studying the ecological behavior of contemporary cities.[2] Vital to the study of ecologies, is having an understanding of the multitude of flow systems that interact with an environment at any given time.

DEFINING FLOW

> Fluids travel easily. They 'flow,' 'spill,' 'run out,' 'splash,' 'pour over,' 'leak,' 'flood,' 'spray,' 'drip,' 'seep,' 'ooze'; unlike solids, they are not easily stopped—they pass around some obstacles, dissolve some others and bore or soak their way through others still.[3]

Human developed landscapes, whether urban or rural, are subject to numerous flows including those of information, people, energy, goods, food, waste, water, and wind. These occur as part of the "natural" systems of the planet (weather patterns, hydrology, solar energy distribution, etc.) and also within the technical systems created by humans (shipping routes, roads, communications networks, etc.). Functioning, or dynamically balanced, ecological systems integrate the complex flows of numerous elements across, under, and above the surface of the earth. Flows can be smooth and continuous, or "saltatory" or interrupted. Ecologies combine both, flows

---

[1] Rob H.G. Jongman and Gloria Pungetti, eds., *Ecological Networks and Greenways: Concept, Design, Implementation* (Cambridge: Cambridge University Press, 2004), p. 5.
[2] See, for example, Brian McGrath and Victoria Marshall, eds., *Designing Patch Dynamics* (New York: GSAPP/Columbia University, 2007). See, also, Stan Allen, "Urbanisms in the Plural," in *Practice: Architecture, Technique and Representation* (Abingdon: Routledge, 2009), pp. 159-191.
[3] Z. Bauman, *Liquid Modernity* (Cambridge: Polity Press, 2000), p. 3.

seeking smoothness, but also accommodating interrupted and inefficient flows, which cause more interactions with environments.

Environments modified by humans tend to segregate and channelize flows in engineered infrastructure systems (pipes, roads, cables, ditches, etc.,) that prevent integrated and balanced flows across various ecological systems. This separation between natural and human managed flows can result in ecological stress; for example, the dumping of all kinds of waste in landfills often removes vital resources from environments. In and around cities, human constructed infrastructure can result in ineffectiveness, inefficiency, and even ecological collapse. Ultimately, integrating flows in urban and agricultural environments are vital to supporting effective ecologies.

Flows are associated with liquids and gases, which show fluidity and change shape according to that which contains a flow. Flows are studied in disciplines such as fluid mechanics (or dynamics), thermodynamics (the flows of heat and work), meteorology (weather patterns), and hydrodynamics (and hydrology). The behavior of liquids is affected by properties such as density, specific weight, viscosity, velocity (rate of flow), depth, turbulence, compressibility, pressure, consistency, surface tension, and degree of confinement. Flows behave in various ways depending on composition and medium, these can be open or channelized, and depending on the factors at play can be smooth or laminar flows or flows subject to varying degrees of turbulence.

Among the specific characteristics of fluids are those of viscosity, velocity, turbulence, and degree of confinement (or resistance to leakage). The viscosity of fluid depends on its internal resistance to flow and is a measure of "thickness" or internal friction; this also encompasses velocities between layers in a flow. The overall speed of a particular flow addresses time and involves the characteristics of the flow itself, and the medium within which a substance is flowing. When there is a disturbance in a flow, typically turbulence is created, or a chaotic type of flow. Turbulence creates eddies, vortices, whorls, whirlpools, and the like.[4] A smooth flow is a laminar flow in which the streamlines are even and predictable, whereas turbulent flows contain numerous eddies. For example, a river typically has both smooth and turbulent flows over the course of its length. Within any system there is loss, or leakage. A leak refers to the escape of flows through a hole or break, often a breach or defect in a system resulting in the loss of material.[5] Typically this involves a boundary, barrier, or wall, which has been perforated, it can be a leaking vessel or pipe, seepage into a substrate, the uncontrolled flow of refugees across a border, or waste in a system.

In a "natural" ecosystem, which is typically dynamically balanced, many flows are operating; these are subject to blockages, disruptions, varying media, changing directions, and interactions with other flows. These effect the smoothness or turbulence of the flow, and how a particular flow interacts with other flowing elements, and with static factors as well. The degree to which a particular flow

---

[4] See P. Ball, *Flow* (Oxford: Oxford University Press, 2009).
[5] *Shorter Oxford English Dictionary* (Oxford: Oxford University Press, 2007), p. 1567.

interacts with other flows and elements, is a function of its composition, velocity, volume, etc.

In examining flows in landscape systems, landscape ecologists Richard T.T. Forman and Michel Godron identify three types: airborne (supersurface), overland (surface), and soil (or subsurface).[6] The various vectors (forces) that drive flows include wind, water, and various organisms (including humans).[7] For example, airflows have much to do with distribution of solar energy. On the surface of the earth animals often move between eco-systems, participating in flows across boundaries. Subsurface flows are usually driven by water movements, many of the toxins produced by humans are moved across surfaces or underground by water flows. The forces moving flows in a system are mainly determined by energy systems, which involve diffusion, mass flow, or locomotion.[8] Diffusion involves dissolved or suspended materials moving from an area of higher concentration to one of lower concentration, mass flows involve "the movement of matter along an energy gradient," and locomotion involves transportation by animals and humans.[9]

Flows within a system are many and varied and can be operating in opposite directions and at different levels.[10] Further, flows can be non-directional and dispersed, or highly directed or channelized. Channelized systems include natural corridor systems such as rivers, and human constructed systems such as pipes and cables, all of which carry flows, usually in one direction. The flows over, across, or under a landscape are measured by the "resistance" of the landscape, which is determined by boundary structures, the "hospitableness" of a landscape to a particular flow, and the length of landscape elements at play.[11]

The flows of liquids, including the movements of gaseous elements, are highly complex and difficult to predict, this is evident in the global and seasonal wind patterns that effect climate and weather. In air movements energy, sound, gases, organisms, and particulates are moved about from environment to environment. In wind flows, the boundary layer, or that layer closest to the surface of the earth is greatly modified by topography, vegetation, and structures such as buildings. In urban cores tall buildings can dramatically influence wind effects, including velocity, intensity, and direction. Another vital flow involves large global cycles of water movement that includes evaporation, precipitation, and various supersurface, surface, and subsurface flows. Much of this water, whether from precipitation directly and surface runoff, from interflows just below the surface, or from groundwater, eventually flows into streams and rivers, where it then moves to oceans. There are many factors that affect the flows of water towards a stream or river, and within the stream or river itself. These include interruptions and breaks, and the system also provides temporary storage (in plants, lakes, groundwater, man-made reservoirs, etc.) that effect overall flows. Underground water flows carry many substances and are

---

[6] R.T.T. Forman and M. Godron, *Landscape Ecology* (New York; John Wiley & Sons, 1986), p. 315.
[7] Ibid., p.315.
[8] Ibid., p. 316.
[9] Ibid., p. 316.
[10] Ibid., p. 320.
[11] Ibid., p. 405.

driven by gravity and soil porosity, flow rates are determined by the volume of flow, the porosity and structure of the soil, and the "filtration effect" of the soil.[12] Surface flows of water can be highly erosive depending on the surface composition and the volume of water flow. Urban hydrology is a vital aspect of the overall ecological performance of a city.

HUMAN CONSTRUCTED FLOW SYSTEMS

Crowds, vehicles, water, air, energy, capital, and goods all operate in flows, or like fluids. These tend to function in mono-functional networked infrastructure that is relatively segregated from other environmental elements. Contemporary cities, as heavily managed ecologies, comprise numerous edges and boundaries between elements and systems, these impede effective flows. Effective flow patterns between the natural and human constructed systems can be negligible as human infrastructure tends to operate independently. Typically, a resource is extracted from a natural environment, processed, used, and then the waste is dumped back into the natural system with little or no regard for how this effects overall ecologies. Humans tend to produce waste, whereas a functioning eco-system does not. An eco-system produces bio-mass, which is then broken down by decomposers in a dynamically stable system. In unbalanced eco-systems, such as an aggrading system where more biomass is produced than can be decomposed, and in a degrading system where there is the inverse, eventually the system will readjust and to move towards stability.[13]

John Urry raises the concept of global fluids, operating in networks and across "scapes."[14] He writes that global fluids

> travel along these various scapes, but they may escape, rather like white blood corpuscles, through the 'wall' into surrounding matter and effect unpredictable consequences upon that matter. Fluids move according to certain novel shapes and temporalities as they break free from the linear, clock time of existing scapes.[15]

This fluidic model addresses contemporary complexity, and the diverse behaviors of the numerous flows that interact in environments today: people flows, information flows, material flows, monetary flows, climate flows, oceanic flows, etc. Flow systems include the extensive global transportation, pipe, cable, and duct systems that move people, goods, water, air, heat, waste, information, etc. For example, in Los Angeles the freeway system handles tens of thousands of vehicles a day in relatively smooth flows, interrupted regularly by accidents, blockages, construction delays, etc. The drivers involved in this flow system obey both written and unwritten rules that

---

[12] Ibid., pp. 335-337.
[13] Ibid., p. 264.
[14] Arjun Appardurai in his book *Modernity at Large* (Minneapolis: University of Minnesota Press, 1996) employs the notion of "scapes."
[15] J. Urry, *Global Complexity* (Cambridge: Polity Press, 2003), p. 60.

allow for continuous self-regulation of the system.[16] The freeways, like the streets of traditional cities, handle flows of people, goods, information, and waste and are vital to the overall functioning of the urban ecology. These local systems are harmonized to global flow systems that both supply and are supplied by a particular city.

In his book *The Informational City*, Manuel Castells examines the impact of information technologies on urban and regional structures leading to his concept of the "space of flows." With its focus on innovation and process, information technologies have reshaped the world in the last few decades. Discussing organizational structures Castells writes:

> The space of flows among units of the organization and among different organizational units is the most significant space for the functioning, the performance, and ultimately, the very existence of every organization. The space of organizations in the informational economy is increasingly a space of flows.[17]

Castells argues that the space of flows, which has its own structure, created by the explosion of information technologies in the last few years has radically changed modes of production, economic and capital systems, organizational structures, spatialities, and ways of communicating. This system of flows is fluidic and multi-directional, it is also placeless, challenging the social importance of cities and their regions.[18] Like flows through an eco-system, effectiveness is vital. And while there can be blockages and disruptions in the flow patterns, these eventually are smoothed out.

Production and economic systems have attempted to develop methods of analysis to make flows more efficient and effective. The famous Toyota Production System (TPS), developed by Taiichi Ohno, was designed to optimize the production of automobiles, by reducing waste in overproduction, time lost, transportation, manufacturing processes, stock, movement, and defects.[19] By reorganizing its production systems and implementing flow mechanisms Toyota developed a more efficient and effective manufacturing model with higher quality, productivity, flexibility, and worker morale.[20] This lean system aims at "just-in-time" delivery, by creating a flow model that is reminiscent of the flows in ecologies.

Another related method used to examine flow systems is Materials Flow Analysis, which attempts to track the use of resources, from extraction, through

---

[16] See Reyner Banham's description of the freeways in *Los Angeles: The Architecture of Four Ecologies* (Harmondsworth: Penguin Books, 1971), pp. 213-222; and P. Ball's description of self-organizing systems in *Flow*, pp. 124-163.
[17] M. Castells, *The Informational City: Information Technology, Economic Restructuring, and the Urban-Regional Process* (Oxford: Basil Blackwell, 1989), p. 169.
[18] Ibid., p. 350.
[19] T. Ohno, *Toyota Production System: Beyond Large-Scale Production* (Cambridge, Mass.: Productivity Press, 1988), pp. 19-20.
[20] See J.K. Liker, *The Toyota Way: 14 Management Principles from the World's Greatest Manufacturer* (New York: McGraw-Hill, 2004), pp. 87-103.

manufacture, consumption, use, and disposal (life cycle analysis).[21] This method both examines material flows as part of economic performance, but also looks for wastage in particular aspects of an economy. As William McDonough and Michael Braungart point out the conventional methods of industrial production tend to be linear and very wasteful.[22] There are significant efficiencies that can be achieved in the flows of materials through the systematic elimination of waste. This is also defined as "hidden flows," those aspects of economic production that include wasteful extraction, manufacturing, and distribution procedures, this also includes processes that environmentally detrimental. Contemporary industrial processes tend to rely on outdated flow methods.

Capitalism is heavily reliant on the continuous and open flow of capital. In *Anti-Oedipus: Capitalism and Schizophrenia* Gilles Deleuze and Félix Guattari address questions of capitalism and flow. Elsewhere, Deleuze states, capitalism "has always been, and still is a remarkable desiring-machine. Flows of money, flows of the means of production, flows of man-power, flows of new markets: it's all desire in flux."[23] In Deleuze and Guattari's work, their concept of flow is related to their concept of coding, decoding (deterritorialization), recoding (reterritorialization), and overcoding.[24] In other words, it is integral to their concept of the assemblage. As Deleuze notes capitalism decodes all flows and is continuously expanding. It cannot create codes, beyond primitive accounting methods, and it ignores limits. As Deleuze says, the "system is leaking all over the place."[25] Capitalism, as it over codes creates surpluses, or leakages. This is evident in efforts to regulate worldwide capital flows. This is also evident in the infrastructure, whether it is leaking pipes, leaking boundaries and borders, information leaks, etc. Inefficiencies in a system are a form of leakage, and while these "hidden flows" (black markets) are costly in a production or distribution system, they can be productive in an ecology or economy. As Henri Lefebvre states:

> The economy may be defined, practically speaking, as the linkage between flows and networks, a linkage guaranteed in a more or less rational way by institutions and programmed to work within the spatial framework where these institutions exercise operational influence. Each flow is of course defined by its origin, its endpoint, and its path. But, while it may thus be defined separately, a flow is only effective to the extent that it enters into relationship with others; the use of energy flow, for instance, is meaningless

---

[21] See D. Rogich, A. Cassara, I. Wernick and M. Miranda, *Material Flows in the United States* (Washington: World Resources Institute, 2008).
[22] See W. McDonough and M. Braungart, *Cradle to Cradle* (New York: North Point Press, 2002).
[23] G. Deleuze, "On Capitalism and Desire," in D. Lapoujade, ed., *Desert Islands and Other Texts 1953-1974* (Los Angeles: Semiotext(e), 2004), p. 267.
[24] R. Due, *Deleuze* (Cambridge: Polity Press, 2007), pp. 65-66.
[25] Deleuze, "On Capitalism and Desire," p. 270.

without a corresponding flow of raw materials. The coordination of such flows occurs within a space.[26]

Lefebvre underscores the complex question regarding the inter-relationship between broad flow systems and network systems. The conflicts that arise between flows and networks are central to the ecology of cities.

Cities have always depended on flows for existence, as Deleuze and Guattari state, the "town is the correlate of the road. The town exists only as a function of circulation, and of circuits; it is a remarkable point on the circuits that create it, and which it creates. It is defined by entries and exits; something must enter it and exit it."[27] Effectively, towns and cities regulate flows, flows from the immediate region, and from far afield, merge in the city where they are then re-distributed. These tend to rely on inefficient infrastructure systems and, as has been demonstrated by Materials Flow Analysis, are uni-directional and highly wasteful. This includes flow inefficiencies in channel systems (roads, aqueducts, communications systems, etc.), which result in leakages of energy and material (hidden or unaccounted flows), and an inherent inability to mesh human created flow systems with "natural" flow systems.

The structure of a city will determine patterns for many types of flow. Human managed flows are often sourced from natural flows, or stored flows, and detoured through human systems, in theory being returned to the eco-system. For example, this is the case with urban water supply which is extracted from regional water sources, primarily rivers and aquifers, processed, consumed, and sent into a sewage system, a storm water system, or the groundwater; waste water going into a sewage system may be processed, and then returned to the natural system. There is leakage or waste in the system, and the processing of water often leads to environmental degradation. In some cases, large quantities of water do not return to the eco-system, as is the case for water used in agriculture and various industries.

As fluidic, flows are dynamic and are subject to various internal and external factors that determine movement patterns. These include the composition of the flow and the environment within which it is moving (permeable or impermeable, contained or uncontained, etc.). Inevitably flows are subject to blockages or disruptions, this can be bends in a pipe, changes in geological structure, depth differences in a river, or changing heat effects in weather patterns. The fluidity of flows, and the interactions that a flow will have with an environment, including other flow systems, is vital to an ecology whether natural or human designed; inevitably, flows transport energy, nutrients, information, goods, etc. Drawing from landscape ecology, environmental elements can act as a habitat, filters, conduits, sources, and/or sinks.[28] Flows can act as habitats for those organisms that either exist within a flow, or for those transported by a flow. The filter and conduit function refer to the degree to which a flow passes through an environment, and how interactions are made. This has to do with

---

[26] H. Lefebvre, *The Production of Space* (Oxford: Blackwell, 1991), p. 347.
[27] G. Deleuze and F. Guattari, *A Thousand Plateaus: Capitalism and Schizophrenia* (Minneapolis: University of Minnesota Press, 1987), p. 432.
[28] See, Richard T.T. Forman, *Land Mosaics: The Ecology of Landscapes and Regions* (Cambridge: Cambridge University Press, 1995).

containment (open or channelized), and boundary systems that either prevent or control flows (degree of leakage). Flows typically cross landscape elements, transporting energy, nutrients, organisms, and waste between eco-systems; these can be vital to the functioning of differing eco-systems, or can be harmful if it is an invasive flow. An environment will also act as a source for flows and be a sink for the storage of flow elements.

TURBULENCE AND HABITATS

Flows can be smooth or laminar, or subject to varying degrees of turbulence; in all flow systems there is bound to be turbulence. Turbulence indicates instability in a flow system, and can be generated by factors within the flow, such as volume, speed, and viscosity, or factors outside of the flow such as the shape of the channel, changes in direction, and obstacles; this is very evident in the flow of a river. Turbulence involves the loss of energy from a flow and is typically seen as a sign of inefficiency. However, turbulence also has beneficial effects and significantly shapes the habitats of numerous organisms, including humans. How do organisms maintain habitat in turbulent environments?

Steven Vogel, in his book *Life in Moving Fluids*, contrasts laminar flow in which "all fluid articles move more or less parallel to each other in smooth paths," against turbulent flow in which "fluid particles move in a highly irregular manner even if the fluid as a whole is traveling in a single direction."[29] Turbulent flows contain numerous eddies, or "regions of swirling flow," and is generated by "infinitesimal disturbances to the flow" that can grow exponentially.[30] Mathieu and Scott note that "turbulence flow is rotational, that is, it contains vorticity."[31] Turbulent flows are defined by randomness, they involve energy transfers, typically originate from instability between laminar layers in a flow, and are "characterized by high levels of fluctuating vorticity."[32] Turbulence dissipates energy, through a cascading of energy loss from larger to smaller scale vortices.[33] The behavior of fluids is effected by a wide range of properties as noted above, many of these determine whether turbulence will be present, and how it will perform. For example, viscosity determines how easily a liquid flow, how prone it is to break up into vortices, and "how steep will be the velocity gradients."[34] Predicting the behavior of turbulent forces is one of most difficult aspects of fluid mechanics.

Many organisms exist within flow systems, including maintaining habitats in turbulent flows. Species that occupy such habitats permanently have long since

---

[29] S. Vogel, *Life in Moving Fluids: The Physical Biology of Flow* (Princeton: Princeton University Press, 1983), pp. 37-38.
[30] See J.A. Fay, *Introduction to Fluid Mechanics* (Cambridge, Mass.: MIT Press, 1998), p. 357.
[31] J. Mathieu and J. Scott, *An Introduction to Turbulent Flow* (Cambridge, UK: Cambridge University Press, 2000), p. 12.
[32] H. Tennekes, and J.L. Lumley, *A First Course in Turbulence* (Cambridge, Mass.: MIT Press, 2001), pp. 1-3.
[33] Ibid., p. 14.
[34] Vogel, *Life in Moving Fluids*, p. 19.

adapted to the vortical, swirling, and cascading aspects of turbulent habitats, whereas those that inhabit these environments temporarily develop a range of provisional adaptive strategies, including those that may involve a significant expenditure of energy, and various ad hoc approaches. If we examine the behavior of trout in a river, we can make the following observations. Turbulence in a river is produced by both the flow of water in the river channel, changes in the shape, depth, and direction of the channel, and obstacles, such as rocks. There is a tendency for river fish to avoid as much as possible extremely turbulent conditions, however, trout are adapted to exist in these environments and, therefore, are also able to use turbulence to their benefit. They have evolved shape, fin performance, eating patterns, and swimming mechanics to live in rivers, they also able to continuously adjust to general flow patterns, and to constant changes in those patterns. Those organisms that have adapted to turbulent conditions typically benefit from the effect of "vortex capture," where energy is harnessed by the organism from the turbulent flows in order to maintain stability. This occurs, for example, when a trout is seeking refuge from the general flow system by swimming in the slower turbulence behind a rock ("station holding").[35] By capturing the energy from turbulent flows, and continuously adjusting to the flows in the environment, organisms, such as fish and birds, are able to achieve a dynamic balance with a habitat.

In the case of humans and the environments they create and manage, Jean Baudrillard points out in his book *The System of Objects*, objects employed in everyday life involve forms of energy; initially human, then animal, and now mechanical and electronic energies.[36] As he writes, in traditional, or pre-industrial settings,

> Man's profound gestural relationship to objects epitomizes his integration into the world, into social structures. We cannot help but admire scythes, baskets, pitchers or ploughs, amalgams of gestures and forces, of symbols and functions, decorated and stylized by human energy and shaped by the forms of the human body, by the exertions they imply and by the matter they transform.[37]

The objects we employ in the creation of habitat help negotiate our spatial and social relationships, these involve continuous energy transfers, much like that of a trout in a river. With the concept of vortex capture, there is a continuous transferring of energy between environment, objects, and humans in the creation of habitats.

As suggested above, living involves the expenditure of energy, or participating in energy flows. Traditionally the home in settled communities, as a fixed immobile construct, was precisely constructed using vernacular practices to respond to a range of localized flows, however, since the advent of modern systems flows are now

---

[35] See J.C. Liao, "A Review of Fish Swimming Mechanics and Behavior in Altered Flows," *Philosophical Transactions: Biological Sciences*, Vol. 362, No. 1487 (Nov. 29, 2007), pp. 1973-1993.
[36] J. Baudrillard, *The System of Objects* (London: Verso, 1996), p. 48.
[37] Ibid., p. 48.

controlled by an array of mechanical and electrical systems. Flows in a home are adjusted by the control of openings (doors and windows), the regulation of machines (thermostats, switches, etc.), and the arrangement of furniture within the spaces of the home; these contribute to the patterns of living in a home. Henri Lefebvre writes:

> our house would emerge as permeated from every direction by streams of energy which run in and out of it by every imaginable route: water, gas, electricity, telephone lines, radio and television signals, and so on. Its image of immobility would be replaced by an image of a complex of mobilities, a nexus of in and out conduits. The occupants of the house perceive, receive and manipulate the energies which the house itself consumes on a massive scale.[38]

Reeves suggests that bad living habits involve the wasting of energy, and bad furniture creates bad habits, in a home "there should only be things designed either to save energy or to allow the good use of it."[39] Furniture responds to habits, or patterns of living, Reeves implies that there are efficient and inefficient (i.e. turbulent) modes of living, but is efficiency a key determinant?

Activities in a home are wide-ranging involving differing types of energy, these include: maintaining, creating, sleeping, bathing, dressing, cooking, eating, playing, child training, recreation, etc.[40] Each of these involves energy expenditure and capture, spaces, furniture, and appliances, and are linked together formally and with the social structure of the home, can be more or less efficient, and are essential to the functioning of the home. Against the many flows that a home participates in the shelter and its contents both prevent and create turbulence. The flows in a home create both harmonious and turbulent conditions, the arranging of objects and domestic space is a continuous operation, the continuous creation of "home"[41]; domestic actions participate in a wide range of flow patterns and continuous exchanges of energy.[42] J. Macgregor Wise has written about the creation of home, or habitat, as the continuous "organization of markers (objects) and the formation of space. Home can be a collection of objects, furniture, and so on that one carries with one from move to move."[43] He suggests the arranging of objects, the organizing of space, and the presence of family and friends, is part of the identity making associated with home. As Wise states, the creation of home is a continuous activity that occurs everywhere we go, through "the arrangement of objects, practices, feelings and affects."[44] Habitat making is habitual, and always endures a certain degree of turbulence.

---

[38] Lefebvre, *The Production of Space*, p. 93.
[39] D. Reeves, *Furniture: An Explanatory History* (London: Faber and Faber, 1948), p. 24.
[40] See R. Woods Kennedy, *The House, and The Art of its Design* (New York: Reinhold Pub. Corp., 1953), pp. 132-133.
[41] M. Csikszentmihalyi and E. Rochberg-Halton, *The meaning of things: Domestic symbols and the self* (Cambridge: Cambridge University Press, 1981), p. 86.
[42] Ibid., p. 195.
[43] J. Macgregor Wise, "Home: Territory and Identity," in *Cultural Studies* 14 (2), (2000), p. 299.
[44] J. MacGregor Wise, "Assemblage," in C. Stivale, ed., *Deleuze: Key Concepts* (Montreal: McGill-Queen's Press,) p. 79.

Nomads and transients space tend to employ human powered implements, whereas settled peoples have developed various labor-saving technologies that change energy usages. This has resulted in a less active relationship with objects that shape habitats. However, these new forms of energy and technologies have led to new gestures, operations, and functions.[45] The homeless tend to employ nomadic, or traditional, technologies that engage the body, although they are typically immersed in new technologies and new forms of flow. Part of existing in turbulent environments is seeking those areas that are less turbulent, or free of turbulence. Therefore, we can suggest that to a large extent the creation of habitat involves transfers of energy between flow systems and the space of a habitat. For nomads, or the homeless, this involves continuous movement and creation of territoriality, or negotiating turbulent flows such wind, animal movements, climate, and encampments. For those living in cities, this is a more controlled relationship, as urban societies attempt to eliminate turbulence from habitats.

The urban homeless are refugees from the city, the homeless are defined by the dominant society and are those whose existence is not seen as conventional, as representing a failed subclass. Transient, and homeless, urban populations are continuously buffeted by the forces of turbulence in cities, however, they also can benefit from inefficiencies in flow systems. The quasi-nomadic aspect of the homeless is evident in the continuous quest for shelter, whether constructed or found in institutions. The scrounging for food and clothing, battles with addictions and mental illness, fighting against forces in the city including the police and other street people, and negotiating with authorities, all represent efforts to adjust and adapt to turbulence. In order to survive, the homeless engage in various forms of "material survival strategies" that involve the expenditure of energy, such as working for street agencies, day labor, peddling, panhandling, prostitution, theft, street performing, and scavenging.[46] The homeless adapt to the turbulence of the city by hiding (camouflage, etc.), by moving constantly through the less turbulent zones (back alleys, marginal spaces), negotiating with authorities, and by adopting nomadic approaches to sourcing heat, clothing, food, and shelter. The homeless operate as individuals, in small groups, and as members of large semi-permanent settlements. Examples of substantial urban encampments include London's infamous "cardboard city" in the 1980s and 1990s, New York City's homeless camps in spaces such as Tompkins Square Park in the 1980s, and Sacramento's "tent city" of the 2000s. For the "homeless" the construction of shelter, the organization of movement, avoiding conflict, and keeping fit involve many strategies including the careful employment of limited personal possessions.

Turbulence has been used to describe the flows of migrants to and across borders and has been defined as "the unsettling effect of an unexpected force that alters your course of movement."[47] In this case, turbulence refers to situations and behaviors outside of a norm, where a system has seemingly broken down. The chaos that

---

[45] Baudrillard, *The System of Objects*, p. 50.
[46] D. Levinson and M. Ross, eds., *Homelessness Handbook* (Great Barrington: Berkshire Publishing, 2007), p. 144.
[47] N. Papastergiadis, *The Turbulence of Migration: Globalization, Deterritorialization and Hybridity* (Cambridge, UK: Polity Press, 2000), p. 4.

ensues, following a political revolution or civil war, leads to numerous types of social, political, and economic turbulence. The flows of refugees fleeing from conflict is another example, where many systems are challenged including the role and defense of borders, the mobilization of international relief efforts, constructing camps, and often dealing with armed conflict. Here, multitudes of people flow often create turbulence at borders and crossing points, particularly where a flow of refugees is blocked at a checkpoint.[48] Refugees, who are able to escape their home country, often end up in the chaotic conditions typically found in refugee camps. The refugee on the road overburdened with possessions, and the makeshift tent in the refugee camp have become symbols of habitat seeking within turbulent flow systems. In all of these cases provisional and adaptive behaviors, institutions, and systems arise to cope with conditions.

Conclusion: Productive Flows

James N. Rosenau, in his book *Turbulence in World Politics*, has produced a useful definition of turbulent activity that crosses disciplines, when he writes that turbulence "sets in when numerous micro actions culminate in macro outcomes that lie outside the system's normal functioning." [49] He also writes that when "the system's boundaries no longer contain the fluctuations of the variables, however, anomalies arise and irregularities set in as structures waver, new processes evolve, outcomes become transitory, and the system enters a period of prolonged disequilibrium."[50] This statement suggests the productive potential in turbulence. The cascading effects of turbulence, as energy is transferred, occurs when a river changes its course, when rapid changes occurs in an organization, or in the construction of both temporary and permanent shelter. The effects of turbulence can result in new patterns, behaviors, structures, and processes, or new forms of operationalizing habitat. Given that contemporary environments, are subjected to high degrees of turbulence within flow systems, the harnessing of energy in habitat creation is vital.

The continuous transformation of a landscape by urbanization results in new ecologies. The ecological productivity of a landscape depends upon the integrated interconnectedness of all the elements in a system. In cities, many aspects of a complex urban ecology are engineered, not allowing for the integration of or harnessing of flows. In order to become more ecologically sustainable, cities must begin to induce flows across their entireties, in other words, across ecological systems. An example of this is the standard handling of storm water run-off in many urban environments in which rainwater is channeled into storm sewers which lead into rivers and oceans, bypassing a wealth of ecological opportunities. Typically, cities segregate urban elements physically and functionally. In a preliminary way, we can suggest that cities need to become more multi-functional and spatially dynamic,

---

[48] Ibid., p. 4.
[49] J.R. Rosenau, *Turbulence in World Politics: A Theory of Change and Continuity* (Princeton: Princeton University Press, 1990), p. 60.
[50] Ibid., p. 8.

or the patches and boundaries that tend to comprise a city need to be better interconnected to be more sustainable. In theory, this means that the impact of requirements for resources and energy, and the production of waste, could be mitigated at the local level, by creating environments that are better integrated with local and regional ecological systems.

Ecologies of the Early Garden City

CHAPTER 2

Patches and Patterns

The flows that traverse environments interact with materials and structure on numerous levels. This chapter will examine the structure of environments drawing largely from theories of landscape ecology developed by Richard T.T. Forman and others.

PATCH-CORRIDOR-MATRIX SPATIAL MODEL

In various works Forman has developed the "patch-corridor-matrix" as a spatial model for understanding the structural behavior of ecologies ranging from wilderness to urban conditions. The overall structure of a landscape (including urban landscapes) describes a "mosaic."[1] In Forman's model the "matrix" describes a general background condition (such as a forest or suburban development), while "corridors" define linear movement elements (such as a river or road system), and "patches" are anomalous and defined territories/spaces (such as a clear cut or a park) in the matrix. This has proven to be a robust and extensively disseminated model.

Patches are defined surfaces and areas that differ from their surroundings, however, they vary widely in their defining characteristics. As a defined piece of land, a patch has a distinctive composition and spatial quality and is produced typically by an intervention in a landscape, either natural or from human action.[2] Forman identifies five types of patches: the "disturbance patch" is created by a localized disruption in a landscape; the "remnant patch" is the inverse of a disturbance patch, it occurs where a piece of a previous landscape survives; the "environmental resource patch" tends to be an anomalous patch in a landscape, or a repeating patchy condition in a more general landscape; the "regenerated" patch is new growth on a disturbed patch; and, the "introduced" patch, such as those found extensively in agricultural and urban landscapes, is the typical result of human activity.[3] Agricultural landscapes made up of fields of differing crops, and urban landscapes made up of blocks of land determined by function, are examples of "patchy" landscapes made from introduced activities.

A second basic system, identified by Forman, is the corridor. Corridors both connect and divide landscapes and are found in landscapes most effected by human interventions. Like patches, they fall into various types: disturbance corridors, remnant corridors, environmental corridors, regenerated corridors, and introduced

---

[1] See Richard T.T. Forman, *Land Mosaics: The Ecology of Landscapes and Regions* (Cambridge: Cambridge University Press, 1995), pp. 3-7.
[2] See Ibid., pp. 43-80, 113-142.
[3] Ibid., pp. 44-45.

corridors.⁴ As Forman stresses, the nature of a corridor depends upon its width, its continuity, nodes or intersections, curvilinearity, connectivity, and other factors.⁵ Like a patchwork system, a corridor system, or network, has a functional effectiveness based on the structure of the system. A mesh of corridor elements creates a network, in which the mesh size and the types of nodes (intersections) define the functioning of the system. A corridor can either be a linear strip removed from a surrounding landscape (such as road through a forest), or a remnant (such as a strip of surviving forest in an agricultural landscape), or it can be a planted condition. In agricultural landscapes, a common corridor typology is the hedgerow or shelterbelt, which modulate wind and energy flows, and can support a diverse range of plant and animal life. As the width of a corridor increases, so does the diversity and complexity of its ecology. The five functions of corridors are: 1) to act as a habitat for various species; 2) to be a conduit for movement; 3) to be a barrier or filter between areas; 4) to act as source; and 5) to function as a sink.⁶ The corridor typology plays a dominant role in defining both the structure and functioning of cities, particularly as conduits of movement, such as found in street systems, and as barriers between adjacent parcels of land. Corridors function in a very similar manner to boundaries, as described in chapter 3.

Thirdly, Forman identifies the existence of a matrix as where one type of landscape element (forest, agricultural fields, urban development, etc.) covers more than half of a landscape, although other factors, such as connectivity, may come into play.⁷ The matrix is the most extensive spatial system and, where present, has the greatest influence on the ecology of a landscape. A landscape matrix has a variety of characteristics, which typically influence connectivity and resistance across a landscape; these determine ecological flows of elements such as energy, water, waste, and organisms.⁸ In an urban or suburban landscape, the matrix may be comprised largely of a network system of corridors and a heterogeneous patchwork of varying landscape types. However, in the case the city, which is, as Forman states, a highly patchy environment,⁹ there is the question as to whether or not the matrix is a useful aspect in analyzing the structure of cities, or should only a "patch-corridor" system be considered? In part this has to scale, or the fineness of urban grain that is being studied. At a more fine scale an urban matrix, becomes a system of patches.¹⁰

When it comes to examining the overall structure and behavior of an environment, or mosaic, landscape ecologists look at relative balance, or imbalance, and/or productivity. The productivity of a given landscape, or eco-system, is measured in terms of Net Primary Production (NPP) or of Net Ecosystem Productivity (NEP), or a measure of the overall activities of producers, consumers, and

---

[4] Ibid., p. 157.
[5] Ibid., pp. 146-147, 153-157.
[6] Ibid., pp. 148-153.
[7] Ibid., p. 277.
[8] Ibid., pp. 285-402.
[9] Ibid., pp. 459-469.
[10] See Forman's discussion of house plots, gardens, and lawns [and buildings], in Richard T.T. Forman, *Urban Ecology* (New York: Cambridge University Press, 2014), pp. 292-306.

decomposers inhabiting an environment.[11] Within the relative balance of an environment, the concept of "resilience" is also employed. An effective system is "dynamically stable" over time, such that while there is continual transformation the overall system does not become unbalanced, or aspects of the system dominate disproportionately to others.[12] In landscape ecology the functioning of a complex landscape is determined by the interactions performance of the overall mosaic as a complex spatial and structural condition. Landscapes can transform quickly, as in the case of a natural disaster, or relatively slowly. The stability of a landscape is dependent on many factors, including resistances in the system and the ability to recover from change (resilience); blockages, porosities, adjacencies, and shape are some of the factors examined in determining a landscape's health and efficiency. Over time landscapes tend to oscillate between stable and unstable conditions, but tend to move towards a dynamic stability, or an ecological balance.[13]

URBAN PATCHWORK DYNAMICS

Human settlement results in the widespread transformation of a landscape, and the establishment of artificial and largely engineered ecologies utilizing complexes of infrastructure. This often results in the disruption of indigenous, or local, ecologies by invasive species, or ones not evolved in a particular locale. Further, the complex ecologies that existed prior to human settlement, or existing in adjacent untouched areas, are usually replaced by the relatively simple (mono-functional) and segregated ecologies created by humans. As noted in the first chapter cities comprise a wide range of flows, which are propelled in many ways, from wind to mechanized transportation; flows can be contained or dispersed, guided, or randomly distributed. Historically, cities have been subject to highly channelized flows which means they are often ecologically unproductive, in that flows are highly controlled, usually prevented from integrating and interacting with broader ecological systems. Cities are also very aggregated landscapes that comprise a high density of patches and corridor systems that have generally evolved over time; cities function as assemblages on a broad spectrum of scales.

Inner cities tend to be rigid environments with a high degree of uniformity in the composition of individual patches of land, resulting in a mosaic at a broad scale, or a complex patchwork of uses at finer scales. The density of street systems in urban environments means that urban blocks are highly segregated from each other, in other words, corridor systems, while comprising communication and distribution networks, are also acting as boundaries. Inner cities are dominated by buildings and hard

---

[11] See C.M. Gough, "Terrestrial Primary Production: Fuel for Life," *Nature Education Knowledge*, 3, (10, 2011), p. 28; J.T. Randerson, F.S. Chapin III, J.W. Harden, J.C. Neff and M.E. Harmon, "Net Ecosystem Production: A Comprehensive Measure of Net Carbon Accumulation of Ecosystems," *Ecological Applications*, Vol. 12, Issue 4, (August 2002), pp. 937-947.
[12] See Douglas G. Sprugel, "Natural Disturbance and Ecosystem Energetics," in S.T.A. Pickett and P.S. White, eds., *The Ecology of Natural Disturbance and Patch Dynamics* (Orlando: Academic Press, Inc., 1985), pp. 344-351.
[13] Forman, *Land Mosaics*, pp. 502-505.

landscapes that restrict the ability to alter the ecological functioning of these areas. Forman and Godron note that suburban landscapes are more diverse, with relatively fewer corridor or matrix elements, where "patchiness" is at a maximum and vegetation diversity is relatively high; greenspace is also abundant, and densities are lower.[14] Therefore, patchwork systems largely defined by transportation networks effectively become the background matrix for the contemporary city. The low or negative ecological productivity in contemporary cities results from a lack of connectivity across very patchy landscapes, where networks of corridor systems and sharp edges prevent crossflows.

We can suggest that suburbia possesses a higher ecological potential due the broader range of patch types and relative openness of the fabric. This phenomenon is reinforced by Brenda Case Scheer who categorizes urban landscape typologies as "static," "elastic," or "campus."[15] The static typology generally reflects inner city block systems. The elastic tissue is generally associated with unplanned and unstable suburban environments, resulting in indeterminate, or "virtually unseen and under-theorized"[16] spaces. The campus typology belongs to planned developments such as airports, apartment developments, and institutions, and is a typology situated between static and elastic tissues. Here, the elastic and campus typologies, ironically the least urban or the least subjected to the subdivision of land, likely hold the greatest ecological potential.

Patches vary widely in their inherent characteristics and can form an overall system that results in a patchwork of defined parcels of land. Cities, as complex ecologies or landscapes, are in constant flux, and are therefore also subject to the various periodicities of daily, seasonal, and annual change. Heterogeneity, diversity, flexibility, flow, and evolution within a landscape are key indicators of ecological functioning. Often the activities of people have the greatest influence on inter-landscape flows, in particular the consumption of energy and non-renewable resources, and the resulting emissions and waste produced, and the effects these have on ecologies. Urbanization has always been about the human management of landscapes, often with little or no understanding of the many and various impacts settlement can have on ecologies, the structure of a given landscape or the elements present, the nature of the flows within the landscape, and changes over time. Ultimately, cities will become more sustainable if the ecological flows across elements in an urban landscape can become more effective, where the management of resources and wastes can be handled in a more integrated, or functionally complex, manner.

Examining in detail the landscape ecology of a city, particularly the operation of patch and corridor systems, is one method for understanding the ecological complexities of urban environments and their associated territories. As highly patchy environments, contemporary cities, aspiring to be more sustainable, will need to work

---

[14] Richard T.T. Forman and Michel Godron, *Landscape Ecology* (New York: John Wiley & Sons, 1986), pp. 302-303.
[15] See Brenda Case Scheer, "The Anatomy of Sprawl," *Places* 14:2 (2001), pp. 28-37.
[16] Albert Pope, *Ladders* (New York: Princeton Architectural Press, 1996), p. 5.

with the inter-relationship between patches. Forman argues that working at the regional level is the best for achieving a sustainable system. He writes:

> A large area in equilibrium that contains many patches in various successional stages has been called a shifting mosaic. Although the total area remains in a steady state, over time patches in different places appear and disappear. In addition to considering shifting mosaic change, patch dynamics focuses on the event or agent causing a patch, and the species changes within it over time. A near-instantaneous disturbance typically is followed by a successional sequence. Each patch exhibits directionality, proceeding from initiation toward 'climax.' The balance between the rate of initiation of patches by disturbance, and the rate of succession within them, determines both the rate and direction of change of the whole mosaic. Hence, the mosaic may be degrading or aggrading, slowly or rapidly, or may be in steady state.[17]

This introduces the concept of "patch dynamics" which is used in landscape ecology as a means for understanding the functioning of landscapes. In highly patchy environments, such as cities, a patchwork system is established which creates a continuously reorganizing mosaic. It is evident that in urban and suburban environments patch and corridor systems are fully interrelated, or one defines the other. While a given patch in a larger urban system is provided its location in a larger corridor network, and is serviced by that network, the interrelationship between the two tends to be rigid and mechanical. The functioning of the overall system depends upon the composition of individual patches, the structure of boundaries (often corridors), and the flows operating within and across the landscape.

Large patches support a more diverse ecology than small ones, however, a sequence of small patches can operate in a similar manner to a single large patch. Patches come in many shapes and sizes, and are caused by a wide range of factors, these include: 1) compact patch shapes, such as a square, conserve energy, whereas patches with convoluted edges enhance interconnections (or flows) with adjacent ecologies; 2) patches function better when they are interconnected with other patches, or they have permeable edges; 3) the relative size of patches and the length of edges (or boundaries), determine how resistant a landscape is to the flow of species, energy, and material.[18] Typically, the edges of a patch support a different ecosystem than its interior. The functioning of a specific patch depends also on its location in an overall system, its immediate adjacencies, and its local context. Therefore, a variety of scales, and the overall aspects of the background matrix influence the behavior of a specific patch. Within a landscape, urban or not, dominated by patches, configurations emerge as flows are established across a system and can take on various patterns.

---

[17] Forman, *Land Mosaics*, p. 44.
[18] Ibid., pp. 113-141.

Evolving Patchwork Patterns

A pattern may be defined as "A regular and intelligible form or sequence discernible in the way in which something happens or is done."[19] It has to do with order or disorder in a system, and whether or not discernible arrangements can be interpreted in that order. The study of urban patterns addresses urban form and structure, land use intensity, the heterogeneity of the structure, and connectivity throughout the system.[20] As has been noted, patterns "are defining characteristics of a system and often, therefore, indicators of essential underlying processes and structures."[21] In the "spatial syntax" theories of Bill Hillier, and others, there is the analysis of morphological settlement patterns, particularly the interconnectivity between spaces and through urban environments. Hillier and Hanson have written that "human spatial organization, whether in the form of settlements or buildings, is the establishment of patterns of relationships composed essentially of boundaries and permeabilities of various kinds."[22] This is consistent with landscape, or urban, ecology.

A patchwork system is a continuously shifting set of alliances, forces, degradations, and aggradations. For example, several patches could form an alignment, or a larger patch. This is similar to the effects created by a patchwork quilt, where patches of cloth sewn together join to make shapes and patterns.[23] Each piece in a patchwork quilt has its own place and characteristics and can exist as a defined area in a larger system or can blend into larger patterns. A patchwork has an underlying order, such as the grid often used by quilt makers, or the street system of a city. A pattern can be a strictly predetermined geometric order, with regularly repeating elements. Or it can be a "crazy," or random pattern. Patterns often involve figures emerging from a ground (or background), they can also be highly repetitive. Pattern can involve the nonrandom elements in a spatial organization such as a city, that which can be predicted.[24] The determination of two-dimensional patterns is dependent on the scale at which it is examined, however, patterns also contain elements operating at differing scales and intensities.[25] Factors such as patch size, composition, density, adjacency, permeability, and patch repetition determine the overall organization of an urban patchwork, and ecological patterns (and processes).

---

[19] *Oxford English Dictionary*, see https://en.oxforddictionaries.com/definition/pattern
[20] See Marina Alberti, "The Effects of Urban Patterns on Ecosystem Function," *International Regional Science Review*, 28:2, (April 2005), p. 175. See also M.G. Turner, "Landscape Ecology: The Effect of Pattern on Process," *Annual Review Ecological Systems*, 20, (1989), pp. 171-197.
[21] V. Grimm, E. Revilla, U. Berger, F. Jeltsch, W.M. Mooij, S.F. Railsback, H.-H. Thulke, J. Weiner, T. Wiegand, and D.L. DeAngelis, "Pattern-Oriented Modeling of Agent-Based Complex Systems: Lessons from Ecology," *Science*, Vol. 310, (November 11, 2005), p. 987.
[22] Bill Hillier and Julienne Hanson, *The Social Logic of Space* (Cambridge: Cambridge University Press, 1984), p. 54.
[23] See Jonathan Holstein, *The Pieced Quilt: An American Design Tradition* (Boston, Mass.: Little, Brown and Co., 1973). For a brief discussion of quilts and patchworks, see also G. Deleuze and F. Guattari, *A Thousand Plateaus: Capitalism and Schizophrenia* (Minneapolis: University of Minnesota Press, 1987), pp. 476-477.
[24] Mark R.T. Dale, *Spatial Pattern Analysis in Plant Ecology* (Cambridge: Cambridge University Press, 1999), p. 12.
[25] Ibid., p. 16.

As Forman and others imply, there are numerous factors both internal and external to a landscape, particularly a patchwork system that can initiate transformations or structural organizations. These can be abetted by human and non-human agents or can emerge from the system as a whole. For example, individual patches, within a larger patchwork structure, can act as countervailing conditions, or they can be a disease-like force that invades, exploits, reverses, or implodes a larger system; this corresponds with the notion of disturbance or planted patches identified by Forman. A system that appears stable, or has been for a long time, can suddenly collapse, or transform. In the contemporary patchwork city, one that is seeking to move towards a continuous space, where ecological flows can be fully integrated, and the pattern of the of the patchwork will be the determining factor, where the size, shape, location, edge conditions, and composition of patches in the larger landscape will be crucial to understanding a system that is dynamically stable. This suggests that patterns within an urban patchwork system would be determined by ecological factors.

Therefore, a city, as a system primarily comprised of patches (legally defined parcels of land with prescribed uses) and corridors, can consist of specific patches or corridors (for example, a park or street). A patchwork system of interacting elements that is constantly realigning as edges open and close. A given patch within the overall structure of a city contains a particular set of organisms, elements, and spaces that are coded according to systems of use and administration. Depending on the interactions within the patch, and its interrelation with a larger system of patches, a given patch will either standalone from larger systems or be active in shifting patterns of performance within the greater patchwork. If a patch is disconnected and mono-functional, as is so often the case in urban environments, then the ecological potential, or its ability to inter-connect, is low. In order for a patch to operate effectively in a larger ecological matrix, the edges must be porous, or a re-functioning of the territory or patch must occur. This is consistent with landscape ecology, which notes that the operations of a patch, and a patchwork, are determined by many factors including size, shape, diversity, and edge conditions.

The breaking down of edges in cities is vital to ecological functionality, as is the creation of movements of flow outside, or against, organized corridors or channels. In the contemporary city, if the background matrix becomes a functioning patchwork with the ability to continuously reorganize itself, then it will begin to behave like a continuous space system with a potential for a high degree of flow of materials, organisms, energy, and waste. Beyond the patch, and patchwork system, is the functioning of the edges or boundaries within a given system. In cities corridors function as both channels for movement, distribution, and communication and barriers between adjacent parcels of land. Cities are often prevented from supporting fluid ecological flows by their extensive corridor networks. One strategy for activating a patchwork system is to selectively reduce or modify the edges in the system, this includes transportation networks, property divisions, zoning regimes, and infrastructure systems. In effect this has been the case in contemporary suburban subdivisions, where there has been a departure from the evenly applied street grid often found in inner cities. Like all complex organizations, cities are subjected to continuous processes of transformation that can modify edges and activate patchworks, these include forces of expansion, collapse, shape changing, migration, etc. These processes of change can modify or divide an existing landscape, through the introduction of holes or cuts, or through fragmentation and decline. Other forces can result in an increased breakdown of a landscape, or the reduction in size or outright loss of elements.

As highly patchy landscapes, cities are also highly fragmented. Many landscapes, particularly urban and suburban, have been subjected to radical fragmentation, which leads to habitat loss, isolation, and ecological breakdown. Patch size, connectivity, and boundary length are determining factors in defining the spatial composition of a landscape.[26] Urban landscapes, as artificial landscapes, once established, transform relatively slowly. Ultimately, cities comprise many types of patch ecologies ranging from small patches of green space, parking lots, buildings, to large urban parks.

ASSEMBLAGES AND TERRITORIES

> The land as the object of agriculture in fact implies a deterritorialization, because instead of people being distributed in an itinerant territory, pieces of land are distributed among people according to a common quantitative criterion (the fertility of plots of equal surface area). That is why the earth, unlike other elements, forms the basis of striation, proceeding by geometry, symmetry, and comparison. The other elements, water, air, wind, and subsoil, cannot be striated and for that reason bear rent only by virtue of their emplacement, in other words, as functions of the land.[27]

Cities and surrounding agricultural lands are "striated" landscapes controlled, or highly regulated, by state organizations. Deleuze and Guattari oppose this condition to "smooth" space, or the open spaces of nomadicism: the prairie, the desert, the sea, etc.[28] In smooth space there are continual processes of deterritorialization at play, with space continuously redefined by cultural systems and movement, here the bounding of space and agency is fluidic. The concept of smooth space resonates with the effective functioning of ecologies, or eco-systems, which strive for connectivity, flow, and dynamic balance. All territories that have been intervened in by humans tend to oscillate between these two conditions, for example, cities have the latent potential to become smooth spaces, or more ecologically effective, by introducing connectivity across systems, through boundary modification, or restructuring the systems themselves.

The territorial operations of assemblages are complex and can be applied to a wide range of spatial structures. Territories, according to Deleuze and Guattari, reorganize functions and regroup forces, a new territory emerges when functions change, and forces realign. Agents of deterritorialization and reterritorialization are constantly in operation. The forces of deterritorialization and reterritorialization are

---

[26] Forman, *Land Mosaics*, pp. 407-412, 426-428.
[27] Deleuze and Guattari, *A Thousand Plateaus*, p. 441.
[28] Ibid., pp. 474-500.

numerous and range from events, affects, flows, exchanges, species interactions, rituals, codes, markings, melodies, performances, etc. As assemblages, houses, gardens, cities, and regions, all emerged from the striation of space and the development of agriculture, urbanization, and military systems. In effect, they are resistant to the forces of smooth space, or the nomadic practices they displaced. And, yet, with its machinic, enunciative, and territorial potential, any assemblage can, and is, transformed continuously.

The concept of assemblage employed by Deleuze and Guattari, with its machinic union of material, expressive, and territorial functions provides a model for further examining the role and function of boundaries in landscapes and organizations. Destabilizing or modifying the boundaries of a territory or organization, is an act of deterriorialization. The resulting, and inevitable, reterritorialization produces a new configuration (arrangement) with a new diagram, and new productive (machinic) potential. These can involve the blurring of spatial and organizational boundaries, leading to a more effective ecological alignment, or can lead towards the hardening of boundaries, or even structural collapse. The interrelationship between a territory, however defined, and as an aspect of an assemblage, and the forces of deterritorialization and reterritorialization are necessary for the spatial definition of the earth, or land. Forces, both internal and external that create these transformations, do so as a special function of the territory, or as a refunctioning of a territory. Specific actions, or agents, can find, define and "assemble" territories, and the forces of deterritorialization and reterritorialization themselves develop new territories.[29] Further, behaviors, interactions, modes of dress, languages, gestural systems, and rituals effect or affect territorialization.[30]

While territories can be limited and relatively fixed (striated), or mobile and unstable (smooth), they can also intermingle, grow or shrink, change shape, be repurposed, be eliminated, or subject to capture. Deleuze and Guattari note that striated and smooth spaces can interchange or can reverse into each other; a striated system can display smooth space tendencies, and a smooth space can become striated.[31] The inversion, or reversal, of a territory, or the emergence of a new assembly, with new machinic-expressive and territorial properties, means that new modes of living can occur, and new territories emerge. Deleuze and Guattari write, "we therefore distinguish between serial, itinerant, or territorial assemblages (which operate by codes) and sedentary, global, or Land assemblages (which operate by overcoding)."[32] Here again they refer to the radically different operational tendencies between smooth and striated territorial systems. However, physical structure is only one aspect of an assemblage (or system of events).[33]

Ultimately, the production of territories is an affirmation of becoming, and the interconnectivity of all events. Territories are fluid organizations participating in

---

[29] Ibid., pp. 326-327.
[30] Ibid., pp. 320-321.
[31] Ibid., p. 474.
[32] Ibid., p. 440.
[33] See M. Bonta and J. Protevi, *Deleuze and Geophilosophy* (Edinburgh: Edinburgh University Press, 2004), pp. 158-159.

forces that shift and transform the structure of a territory. The changing of territories through the processes of deterritorialization and reterritorialization occurs through the interaction of bodies, affects, language, song, and the nomadic line of flight. Nomadicism is also a powerful force that counteracts, disrupts, or destroys territories, it is the agent of deterritorialization; rather than a destabilizing force, it is a productive force, against more rigid, or striated, structures. These forces are stronger or weaker depending on the assemblage, and the degree of smoothness or striation in the context. When an assemblage reaches a threshold, it deterritorializes and reterritorializes which creates new functions, patterns of behavior, machinic potential, species, spaces, codes, etc.

Conclusion: Activating Ecologies

Cities are continuously subjected to forces that belong to material, social, and ecological flows, and are also effected by a wide range of structures, intensities, political and bureaucratic regimes, weather patterns, geological forces, natural disasters, infestations, etc. Herein lies the potential for urban and suburban environments to be rendered more ecologically connected, by allowing cities to accommodate the flows of energy, materials, information, and social systems more effectively. In other words, the entire field of operations can be activated rather than being reliant on independent networks poorly interconnected and unable to provide for the functioning of a total urban system. The patches of land that comprise suburban and urban environments can be reworked, as currently they tend to be inert or have a negative impact. The boundaries that define most urban patches are infrastructure systems that directionally channel flows but tend to be significant barriers to cross flows. The high degree of edges and boundaries in urban landscapes runs counter to the development of effective urban ecologies. There are forces within and without a system that coalesce to create a new alignment, arrangement, or configuration.

In order to improve the connectivity across patchy landscapes, the patchwork needs to become activated, or a shifting system of patterns and changing arrangements must occur. The networks of corridor systems that define urban patchworks provide an important function in cities, however, these also function as edges and boundaries. In other words, corridor systems should also be read as edges, this would describe them as more complex urban structures. The ability for a patchwork to become active depends upon the structure of the edges between individual pieces and the structure of patches in play. To make patch and corridor landscapes of suburbs and cities more connected, the corridors should function as a system acting as part of a distribution network, but also as a system of permeable edges allowing for cross flows, habitats, and sources/sinks. While both suburban and urban landscapes tend to be highly patchy, the suburb supports a functioning patchwork system more readily. The inner city is more resistant because of the relative impermeability of the patches and the high preponderance of rigid corridor networks. Unfortunately, urban patches tend to be inert, as they are functionally determined, and precisely bounded, they tend to exist as disconnected islands; they are the result of urban decision making, in particular the zoning of urban space. In order to generate flow and interconnectivity, the plethora of edges that subdivide cities must be softened. Every patch in a city has the potential to direct movement or flow.

Typically, some human intervention in a landscape increases the heterogeneity of the landscape, however, too much disturbance leads to a homogenization of the ecology of landscape. Both suburban and urban patchworks tend to be ecologically homogeneous or mono-cultural, often comprising imported or invasive species and/or typologies, with sharply defined boundaries that allow for little cross-migration or species interaction. Cities are functions of numerous types of flows, mainly long predetermined routes and channels: the road, the freeway, the flight path, the telephone cable, the sewer line, etc. The control of flows has existed throughout the history of urbanization. Social structures have always operated through and against predetermined conduits, and electronic technologies are also able to operate in more diffused ways.

The emergence of crossflows, triggered by the reorganization of cities, will require the changing of boundaries, the harnessing of forces, and the operations of certain agents (such as urban governments, communities, design professionals, etc.). As we expand the applicability of the powerful concepts of landscape ecology (and patch dynamics) to contemporary cities, we embrace the notion that urban structures are singular (in that they embrace spatial, temporal, social, affective, and material qualities) and infinitely complex. If we consider a city as a very complex set of interlocking ecologies, we can further understand cities as an arrangement of systems, spaces, elements, actions, affects, bodies, languages, structures, codes, etc. Cities encompass a multiplicity of forces that can modify the ecological and spatial structures that are continuously at play. Currently, the heavy reliance on organized channels of movement, functional zoning, and property ownership, prevents the ecological harmonization of contemporary cities. The creation of greater interconnectivity through more complex boundaries is a necessary step to move cities towards sustainability; fully integrated landscapes of patches and corridors will make contemporary cities more ecologically effective.

Borrowing from landscape ecology, we can describe contemporary cities as complex patchworks comprised of segregated parcels of land divided by infrastructure, property laws, and functionality. Patchwork systems depend on the edges between landscape patches, the inter-relationships between patches, and connections between patches and other ecological typologies. Patchworks can be activated in ecological terms by examining the functions of boundaries (corridors) and patches within an overall urban pattern. Converting a striated condition to a smooth space patchwork can be triggered by internal forces that incrementally, and/or inadvertently, lead to change through the modification of edges and boundaries. On the other hand, strategic actions, or interventions, by state or capitalist agencies could lead to the erosion of urban boundaries, and the creation of ecologically interconnected landscapes within contemporary cities. Regardless, this is more likely in the "elastic"[34] landscapes of suburban environments, which are more heterogeneous

---

[34] See Scheer, "The Anatomy of Sprawl," pp. 28-37.

and open, than inner-city landscapes. It is a significant challenge to make the highly interrupted landscapes of cities more interconnected. Ultimately, as in the case of deterrritorializing and reterritorializing aspects of assemblages, this will depend upon the shifting and reorganizing of the numerous boundaries, or edges, that fragment cities in discrete, and often unproductive, ecologies. While urban bureaucracies could trigger, it may be more likely that urban ecologies will be self-organizing, reliant on unexpected boundary changes, shifting ecological alliances, or new diagrams.

CHAPTER 3

Boundaries and Corridors

> It is thus that the frontiers which for a long time were the most impenetrable slowly become transparent; the intermediary regions, the regions of passage, the doors, the interstices become new centers (Hong Kong, Singapore, the Rivieras) toward which crowds converge and form which they spread out, instructed in a new way to listen to things.[1]

Boundaries and corridors systems play a vital role in the structure and behavior of an environment, whether urban or non-urban. Urbanized cultures have a propensity for subdividing land and constructing extensive infrastructure systems, in other words creating boundaries, or division; corridors also participate in networks. This chapter will briefly present this topic, and the similarities in function between boundaries and corridors.

THE FUNCTIONS OF BOUNDARY SYSTEMS

Boundaries typically are narrowly defined linear elements in a landscape. Boundaries have a discernible thickness, actually or virtually, however, regardless of how wide a boundary is it is situated between two (or more) adjoining entities; boundaries define a figure against a field, subdivide systems, inside from outside, and delineate organizational and spatial entities. Boundaries establish directionalities and territorialities, and can generate competition and conflict. A boundary also creates a zone or borderland, typically a narrow territoriality that blends they characteristics of the adjacent entities. Boundary systems imply that there are territories, and territories are typically defined by centers of power. A system of boundaries (defining nations, ecosystems, organizational structures, etc.) is a continuously reorganizing network defining territories, patches, or entities. As relationships between territories in the system change, so does the overall balance in the system. In some places the boundaries are resistant to exchange, while in others they are on the verge of erasure. Flows of economic, social, political, and ecological factors are continuously being redirected by boundary systems.

If we draw from the discipline of landscape ecology, it can be determined that boundary functions are: habitat, filter, conduit, source, and sink.[2] Ideally, a boundary functions as all of these. The boundary is the limit, or the edge, boundaries create a

---

[1] Michel Butor, "Meditation on the Frontier," in *Frontiers* (Birmingham, Alabama: Summa Publications, Inc., 1989), p. 101.
[2] Richard T.T. Forman, *Land Mosaics: The Ecology of Landscapes and Regions* (Cambridge: Cambridge University Press, 1995), p. 96.

zone that is either open or closed to cross movements of flows. Boundaries are also corridors of movement and add and subtract elements from a system. There are particular species and people associated with borders. The functioning of a boundary depends on the adjacent structures, territories, the properties of the boundary, and the type of flows that are moving across or along it, this in turn defines the habitat and what can occupy it. Humans create a multitude of boundary, or edge, conditions through urban and rural settlement patterns. For example, when a boundary is moving outwards, then it is considered a frontier. Those frontier people living on this edge act as an "avant-garde," or a force of expansion, or colonization.

The notion that a boundary is a habitat underscores the territorial functions of boundaries. As a habitat, boundary conditions can support species that are similar to those inhabiting the surrounding areas (patches), but in differing concentrations. This suggests, as noted above, that the boundary is a zone that can support a variety of organisms. Organisms that can occupy the boundary, while belonging to the adjacent territory, can also have distinct characteristics. In some cases, the boundary habitat supports those not found elsewhere in the system, there are a wide range of boundary dwellers who act as part of larger territories or on their own. These marginal species or members of a society, have a special role to play in systems. Boundary dwellers include the "edgewalker" who appears in some aboriginal cultures,[3] and more contemporary figures such as custom agents, marketers, lobbyists, sales people, envoys, mediators, and spies. Often, these edge species are go-betweens, protectors, criminals, outcasts, drop-outs, refugees, monsters, subversives, or those organisms, such as weeds, that have adapted to the boundary condition and exist best in the in-between. Despite the apparent thinness of boundary territories, these are vital habitats, often playing an essential role in the overall functioning of a system.

Boundaries, like membranes, operate with varying degrees of permeability, or porosity: some materials and organisms may cross, and others may be blocked, or partially blocked, depending on many factors.[4] There are elements in the structure of any boundary (or membrane) that can inhibit, facilitate, channel, deflect, or block movements and flows, these can include changes in composition, ruptures, or accumulations of foreign materials.[5] Further, while a boundary may block one kind of flow, it may be completely permeable to numerous other flows.[6] Boundaries, play a role in defining the limits of functional territories, and in regulating the flows of materials, organisms, energy, information, and the like, much of which depends on either "mass flows" or "locomotion."[7] Mass flows refer to movement along an energy gradient, while locomotion involves the transportation of material by animals and vehicles. The porosity of a boundary depends on the composition of the boundary, and what is trying to cross, or pass through the boundary; as a filter a boundary can be

---

[3] See Judi Neal, *Edgewalkers* (Westport, Ct.: Praeger Publishers, 2006).
[4] Forman, *Land Mosaics*, p. 96.
[5] Ibid., p. 97.
[6] J.A. Wiens, C.S. Crawford, and J.R. Gosz, "Boundary dynamics: a conceptual framework for studying landscape ecosystems," in *OIKOS* 45:3, (1985), p. 425.
[7] Forman, *Land Mosaics*, p. 100.

defined as "any device or arrangement which removes or separates out constituents."[8] There are holes in boundaries, either constructed or inherent, that allow for the passage of materials and organisms. In constructed environments these include gates, doors, windows, checkpoints, and the like, all of which modulate flows and movements. The filtering effect of a boundary pertains to the structure of the boundary and the entity supporting it; there are also active agents that can participate in the filtering of various flows across a boundary. The filtering function is the primary role of boundary systems, this establishes how adjacent territories and organizations interact. The stronger the boundaries are in a system, the less integration between bounded entities. Boundary systems tend to be penetrated by holes that create leaks and linkages, however, numerous elements are used to modulate boundary systems.[9]

While a boundary as a filter involves flows that cross, or attempt to cross, a boundary can also function as a channel along its length, often becoming part of a corridor or network system. That which flows along a boundary plays a particular function in a larger territorial system. As Forman and Moore note, boundaries function just like corridors, they write: "Corridors serve as conduits for movement along the corridors, filters of movement across them, sources of effects on their surroundings, sinks for objects in the matrix, and habitats for edge species."[10] A difference between boundaries and corridors, is that the filtering function is primary in boundary systems, whereas corridors function more actively as conduits, or channels, for the movement of materials, information, organisms, energy, and the like. In urban environments, where transportation systems act as a primary element in defining the structure and functioning of cities, they act primarily as networked corridor systems, but they also rigidly subdivide the environment, acting often as impenetrable boundaries to many types of flow. In this case the intersections, or nodes, are also vital parts of the system, with a distinctive intensity. This discussion of boundaries as channels of movement and complex interactive networks, invokes Gilles Deleuze and Félix Guattari's concept of the "rhizome."[11] This also underscores the notion that boundaries are part of extensive networks, and continuities in boundary systems are necessary for them to function. Corridor systems, such as road networks, oscillate between operating as boundaries and preventing cross flows, and acting as conduits for flows; while blocking flows, boundary networks also facilitate rhizomatic flows and interconnectivity.

The final two functions of boundaries, are the concepts of source and sink; a source gives off things and a sink absorb things.[12] As a source, a boundary is the

---

[8] *Shorter Oxford English Dictionary* (Oxford: Oxford University Press, 2007), p. 963.
[9] See W.J. Mitchell, "Boundaries/Networks," in A.K. Sykes, ed., *Constructing a New Agenda: Architectural Theory 1993-2009* (New York: Princeton Architectural Press, 2010), p. 228-230.
[10] R.T.T. Forman and P.N. Moore, "Theoretical Foundations for Understanding Boundaries in Landscape Mosaics," in, A.J. Hansen and F. di Castri, eds., *Landscape Boundaries: Consequences for Biotic Diversity and Ecological Flows* (New York: Springer-Verlag, 1992), p. 247.
[11] See G. Deleuze and F. Guattari, *A Thousand Plateaus: Capitalism and Schizophrenia* (Minneapolis: University of Minnesota Press, 1987), pp. 3-25.
[12] Richard T.T. Forman and Michel Godron, *Landscape Ecology* (New York: John Wiley & Sons, Inc., 1986), p. 315.

location from where something emanates, originates, or flows. This reflects the idea, that the edge or boundary of a territory or organization impacts on both the entire system, but also directly on the adjacent territories directly. Within the structure of a cell in an organism, all elements have a role to play, the same applies to territories and organizations. The edge, or boundary is the zone that interacts with other adjacent structures, it is the point of resistance, conflict, or exchange. If enough accumulates in the boundary zone, then it will eventually give off its own flows, it may become a new territory with its own defined boundaries. This phenomenon is like the processes of deterritorialization and reterritorialization that Deleuze and Guattari describe as being part of an assemblage. According to Forman and Moore, the sink effect "refers to the absorption or accumulation of objects by the boundary."[13] In other words, materials, energy, organisms, and /or information can be purposefully or inadvertently trapped, or stored, in a boundary condition. In physics, a sink is "a place where, or a process by which, energy or some specific component leaves a system."[14] Flows that are do not pass through or are not reflected by a boundary, can be blocked, and hence can accumulate as a sink condition. These can become a significant factor in the composition of a boundary, resulting in its modification, or transformation. In extreme cases these can negate the boundary altogether. As a sink, however, they can also participate in the reorganization of a system. A sink might also provide a means for ameliorating situations elsewhere in a system. When the component that has been removed from the system reenters the system, then the sink switches to a source. As a distinct element in a landscape, or territorial system, boundaries will inevitably both add and subtract components from a larger system. This may be the temporary storage, and release, of solar energy, or the holding of migrants at a border station. The total resources in an environment determine the amount of boundary activity and/or conflict.[15]

BOUNDARY DYNAMICS AND ECOTONES

The structure of a boundary determines how it functions, as do the function of adjacent territories, and various flows across and within a boundary. Further, these many factors (including various "disturbance" factors) can cause the destabilization of a boundary. Like other landscape and organizational elements, boundaries can be formed suddenly or slowly, they can migrate or shift, and they can be breached or erased. Boundaries define territories and organizations, and the type and degree of interaction between entities. Boundaries regulate, or modulate, the flows and exchanges (materials, energy, organisms, or information) in a landscape or between organizations.[16] According to Cadenasso and Pickett, the amount of flow and

---

[13] Forman and Moore, "Theoretical Foundations," p. 247.
[14] *Shorter Oxford English Dictionary*, p. 2845.
[15] H. Aldrich and D. Herker, "Boundary Spanning Roles and Organization Structure," in *The Academy of Management Review*, Vol.2, No. 2 (April 1977), p. 224.
[16] M.L. Cadenasso and S.T.A. Pickett, "Boundaries as Structural and Functional Entities in Landscapes: Understanding Flows in Ecology and Urban Design," in B. McGrath, V. Marshall, M.L. Cadenasso, J. M.

exchange depends on the type of exchange, the composition of the territories, and the structure of the boundary.[17] Boundaries operate as filter mechanisms regulating flows, as part of a boundary network, as habitats supporting organisms, and as a vital source and/or sink for elements in the larger system. In a dynamically balanced system, such as an eco-system or bureaucracy, boundaries play a vital role, and perform all five functions.

Architects have tended to seek stable boundary systems, most notably in the construction of walls that act as boundaries between inside and outside, and between functions. And yet, even in the most rigid of architectural orders there always remains a certain instability in boundary systems. Within environments created by humans (both urban and agricultural), there exists a plethora of boundaries, both on the surface of the earth and in human organizations. These create highly rigid and striated territories that typically are inefficient and highly disrupted. To introduce smoother flows, or to move to a smooth space condition, requires they are operationalizing of territorial and boundary systems in new ways. Within the dynamics of these systems, there are many opportunities for modification, to allow for more effective flows.

The typically impermeable boundaries created by the settlement patterns of urban-agricultural societies tend to create relatively static environments. In nomadic cultures, which seek different forms of stability from urban-agricultural societies, there are seasonal patterns and boundaries at play, the journey tends to be primary; as Deleuze and Guattari note nomad space "is localized and not delimited."[18] In nomadic hunter-gatherer cultures, boundaries are fluid and shifting as groups moved through a landscape, continuously adjusting to topography, sources of food, the seasons, the supernatural, and competing groups.

Boundaries may be physical and appear arbitrary, as in the case of some national borders, or they may be relatively intangible, such as limits of personal space. Many boundaries are cultural constructs than can be inadvertent and often invisible, including the borders between nations. Territories are defined on many levels, and operate on different scales; they overlap, nest, conquer, eliminate, bond, adjoin, invade, etc. Boundaries in urban environments are being continuously made and remade, on varying time frames. Those that appear important at one point in time, can be eliminated, or erased, or softened, or strengthened. One type of flow, such as information, depends on various network and surface systems to function. Information systems will have boundary constructs that include barriers, sources, conduits, and sinks. Another flow type, passing through and across the same field, will have a different system of boundaries.

The qualities of boundaries, both in terms of their relationship to entities and as entities themselves, are a vital aspect of the functioning of an organization. Like all spatial conditions, these are in a constant state of change as forces and flows across and along the boundary vary. The edge of a territory or organization, or the space

---

Grove, S.T.A. Pickett, R, Plunz, and J. Towers, eds., *Designing Patch Dynamics* (New York: GSAPP, Columbia University, 2007), p. 119-120.
[17] Ibid., pp. 120-121.
[18] Deleuze and Guattari, *A Thousand Plateaus*, p. 382.

influenced by a boundary, denotes a zone of influence. This is true of both the boundary as a habitat condition, but also for political, geographical, and organizational boundaries. State systems, corporations, and other societal organizations are governed by codes, authority figures, structural dynamics, and jurisdiction, these determine to a large extent the functioning of boundaries.

Some materials or species will encounter a boundary, and pass directly through it, others will become trapped in the boundary for some length of time (habitat or sink), while others will use the boundary as a secretive movement system. Those using it as a conduit may be using it as form of protection. All of these conditions employ transitional devices, or intermodal exchanges. It is evident that boundaries are an active part of ecological and organizational systems, and that boundaries function in complex ways. Recognizing that boundaries are porous and unstable can make entities function better. In territorial and organization structures the overall functioning of a system is more effective if boundaries have permeability and flexibility. Boundaries, as they filter flows, support various boundary dwellers, and remove and add elements, are an essential aspect of systems; boundaries act in the deterritorialization and reterritorialization of human ecologies.

In every eco-system, there is a direct relationship between the structure of a landscape and the way flows operate within it. Landscapes which are more homogeneous tend to support more continuous flows due to the relative lack of interruptions, whereas flows within landscapes with a multitude of boundaries will depend on the permeability of the boundaries. Edges are influenced by a wide variety of activities, they are habitats in and of themselves, and ecologically different from the adjacent ecologies, as edges often contain a high degree of diversity. Microclimatic conditions, influenced by sun, precipitation, and wind, are very different in edge conditions than from the interior environment of a patch. Through their actions, humans have an enormous impact on the creation and maintenance of edges in a landscape, whether rural, suburban, or urban, humans tend to aggressively maintain edges in the environments they create; these edges are often barriers to ecological flows.

The control of the movements and flows across edges or boundaries are determined by flows of wind and water, or by locomotion (animal, human, and machine), as all edges are filters that contain some degree of permeability. [19] Ecologists, in determining the effectiveness of a landscape examine the resistance to flows inherent to that landscape. According to Forman, landscape resistance is described "as the effect of structural characteristics of a landscape impeding the flow of objects (species, energy, and material). Since boundaries separating spatial elements are locations where objects usually accelerate or slow down, it has been suggested that that boundary-crossing frequency, i.e., the number of boundaries per unit length of route, is a useful measure of resistance."[20] Resistance, is a way of understanding flows in a system and where flows move in an integrated manner, or where they are blocked by impervious barriers.

---

[19] Forman, *Land Mosaics*, p. 100.
[20] Ibid., p. 279.

Cities typically have high degrees of both heterogeneity and homogeneity, and are subject to many sharp, and often impermeable, boundaries that prevent flows outside of predetermined channels. This leads to flows that are saltatory, or subject to interruptions. This concentrates on how connected and smooth flows are, and how they interact with a landscape. However, saltatory flows do also involve more interactions between a landscape and flow of elements.[21] In natural landscapes edges tend to be curvilinear and complex, comprising soft concave or convex surfaces, with a high degree of interdigitation or interaction, and therefore, flows across the boundary. In landscapes significantly modified by human intervention, boundaries tend to be hard, straight, and sharp, hindering flows of organisms, energy, materials, and the like.[22] Therefore, ecological flows across a landscape are a function of the patches, and corridors, but also the types of edges. The soft edge between two ecologies creates a blended condition, or "ecotone," which acts as an inter-system; this occurs extensively in natural landscapes as a transition between eco-systems. In contemporary cities, with its hard edges, the functional separation of urban elements (patches, corridors, infrastructure, etc.) is reinforced by zoning and planning regimes, and is enhanced by the widespread use of the "buffer":

> The concept of a buffer, as an area that lessens or cushions the effect of one area on another, is common in land-use discussions. It relates to the juxtaposition of two areas with incompatible interactions. A buffer separating two areas is sometimes proposed to minimize negative interactions, or to reduce steep gradients in an edge. The buffer may repel, or may absorb, those flows. These are the same two functions accomplished by the many ways of sculpting and managing boundaries.[23]

The buffer technique used in many contemporary cities functionally separates urban elements, and has a theoretical similarity to an ecotone, however, typically does not perform the same role, acting more like an independent element, rather than a thick edge or boundary. In fact, the use of functional city zoning is designed to prevent flows between pieces of land. The softening of buffer systems in contemporary cities is another way that crossflows could be encouraged.

As noted above, the types of edges, particularly within a patchwork or matrix system, are crucial. This indicates that edges can either be simple or complex in the way that they interact with a larger system. Edges, or boundary conditions, depending on width, composition and adjacent patches can enhance or prevent flows, and can also act as a spatial order that supports an eco-system as a part of the management of resources, wastes, and organisms. An edge, on its composition, can be a location for organisms that occupy the inter-zone between ecosystems, a barrier, an open condition that facilitates free movement, and/or a location that generates or stores materials, energy, nutrients, etc. The differentials in permeability caused by varying

---

[21] Forman and Godron, *Landscape Ecology*, pp. 357-361.
[22] Forman, *Land Mosaics*, p. 83.
[23] Ibid., p. 292.

edge conditions within a patchwork, or matrix, determines the degree and types of flow.[24] These characteristics are very similar to the way that corridors function, as defined above. There is a direct inter-relationship between the composition of edges and the functioning of a particular patch; edges have different characteristics and thicknesses based on the size and shape of a patch, orientation and climate, the relative age of elements in the system, the activity of organisms, the overall composition of the patch, and the structure of various adjacencies. Edges and boundaries can advance and retreat over time depending on the forces at play, although human created and maintained boundaries in landscapes are typically quite stable.[25]

Corridors and Networks

Like a boundary, corridors function as habitats, conduits, filters, sources, and sinks.[26] The primary function of corridors is as a conduit for the movement of organisms, goods, information, waste, energy, machines, etc. Secondarily, corridors perform a filtering function (like boundaries) to flows that cross, or attempt to cross, them; corridors often separate patches. In a similar manner to boundaries, corridors also provide habitat for certain species and occupations, and they can be sources and sinks. Forman identifies width and connectivity as the two most important factors in the behavior of corridor systems.[27]

Corridors are often part of a network system that involves a mesh of corridors interconnected by nodes; a network is "a set of interconnected nodes."[28] The functional complexity of a network depends on the arrangement and performance of corridors and nodes (intersections). There are three kinds of networks: 1) "line or chain" networks, where nodes are arranged in a linear fashion; 2) "star of hub" networks where movement occurs through central nodes; and 3) "all-channel" networks where movement occurs in all directions.[29] Networks can be dendritic (arborescent) or mesh-like (rhizomatic).[30] One is more hierarchical than the other. One has controlled directional movements and the other more open multi-directional movements.

The behavior of corridors is in a network is effected by factors such as width, length between nodes (intersections), curvilinearity, porosity of edges, and context. They often interface with other networks. Networks can vary in size, complexity and functionality. Networks can vary in size, complexity and functionality. John Urry writes:

---

[24] Ibid., p. 96.
[25] Ibid., pp. 104-111.
[26] Ibid., pp. 145-149.
[27] Ibid., pp. 153-157.
[28] John Urry, *Global Complexity* (Cambridge: Polity, 2003), p. 9.
[29] Ibid., p. 51.
[30] See Deleuze and Guattari, *A Thousand Plateaus*, pp. 3-25.

> The power of any network can be said to stem from its size, as indicated by the number of nodes in it, by the density of networked connections between each node, and by the connections that the network has with other networks.[31]

Networks rely on the number of nodes, the arrangement of nodes, interconnectivity between nodes, and the intensity of traffic through specific nodes. Nodes can overheat or become dormant. The position of a node in a network is important, as is the ability for networks to interface with other networks. The flow of information, materials, organisms, and the like through networks depends on these factors, including traffic in the system. Flows can be smooth and continuous, or disrupted and blocked. Pathways through networks typically are the shortest route between nodes, but this is not always the case. If there are holes in the network, or blockages, then detours must be taken. Sometimes a detour is faster than the shortest route. Routing through networks is a vital factor in performance (time and/or distance) and productivity.

Increases in connectivity (or number of nodes), lead to increases in the power and productivity of a network. Networks often have hierarchies (and sub-regions) within them, including route hierarchies. Routes can be predetermined and common based on flows and network structure or forged by the flow itself. The potential for varied routes through a network, such as an urban street system, makes the network powerful and also potentially inefficient. In cities many flows follow established routes, often determined by the size and efficiency of internodal corridors. These are prone to disruptions and blockages. Many routes are also mono-functional and poorly interconnected. Urban traffic systems are the clearest example of this. Inordinate effort goes into the "traffic engineering" of cities, but the results, despite the advent of new traffic management technologies, are still susceptible to accidents, traffic volumes, weather, and the like.

Globally integrated networks (GINs) are a contemporary phenomenon that "consist of complex, enduring and predictable networked connections between peoples, objects and technologies stretching across multiple and distant spaces and times."[32] These govern the function of all kinds of entities, and support a broad range of flows. As Manuel Castells states in *The Rise of the Network Society*, networks are very prevalent in contemporary society. He writes:

> Networks are open structures, able to expand without limits, integrating new modes as long as they are able to communicate with the network, namely as long as they share the same communication codes.[33]

Networks tend to be dynamic and open systems that can innovate. They can be hierarchical or not, and can continuously reorganize, especially power relationships.[34]

---

[31] Urry, *Global Complexity*, p. 52.
[32] Ibid., pp. 56-57.
[33] Manuel Castells, *The Rise of the Network Society* (Oxford: Blackwell Publishers, 1996), p. 470.
[34] Ibid., pp. 470-471.

Urban corridor and network patterns can operate at large scales. Most of these are engineered and provide channels for the large volume of flows in cities. With the advent of the information age in the 19th century, many flows also occur through airwaves. A city becomes a complex construct of overlapping and interconnected networks. In his book *Streets & Patterns*, Stephen Marshall has looked at overall urban street patterns, and has determined four basic typologies: a) "altstadt" is based on the irregular and angular street networks of older cities; b) "bilateral" is based on the gridiron structure of certain older cities and more recent cities such as the nineteenth century American city; c) "characteristic/conjoint" which combines both open grid and non-grid elements; and d) "distributory" which reflects the peripheral street arrangement of many contemporary cities by employing hierarchical (arborescent) and cul-de-sac structures.[35] These basic patterns vary from open or rhizomatic, to hierarchical and aborescent.

CONCLUSION: ASSEMBLAGES AND BOUNDARIES

Within highly fragmented environments, such as those created by humans (both urban and agricultural), there exists a plethora of boundaries, both on the surface of the earth and in human organizations. These create highly rigid and striated territories that typically are inefficient and highly disrupted. To introduce smoother flows, or to move to a smooth space condition, requires they are operationalizing of territorial and boundary systems in new ways. Within the dynamics of these systems, there are many opportunities for modification, to allow for more effective flows. Deleuze and Guattari argue that smooth and striated space coexist, and that productive relationships emerge when for those that exist in-between:

> On one side, we have the rigid segmentarity of the Roman Empire, with its center of resonance and periphery, its State, its Pax Romana, its geometry, its camps, its limes (boundary lines). Then, on the horizon, there is an entirely different kind of line, the line of the nomads who come in off the steppes, venture a fluid and active escape, sow deterritorialization everywhere, launch flows whose quanta heat up and are swept along by a Stateless war machine. The migrant barbarians are indeed between the two: they come and go, cross and recross frontiers, pillage and ransom, but also integrate and reterritorialize.[36]

Here, Deleuze and Guattari describe the rigid and imposed boundary systems of state run urban and agricultural societies, the fluid and shifting boundary systems of nomadic cultures focused on the journey, and the hybrid actions of barbaric migrants who act as agents of reterritorilization. They imply that there is a close functional and

---

[35] S. Marshall, *Streets & Patterns* (Abingdon: Spon Press, 2005), pp. 83-89.
[36] M. Bonta and J. Protevi, *Deleuze and Geophilosophy* (Edinburgh: Edinburgh University Press, 2004), p. 222.

ecological relationship between the structure of human organizations and the structure of spatial territorial systems.

Assemblages create and operate with territories that are in constant processes of change: we do not know what an assemblage is until we know how it functions. Functional transformation is characteristic of assemblages, and this aspect has particular importance for buildings and cities. Further, the concept of assemblage suggests that a city can be conceived of as a complex and continuously changing set of interlocking and overlapping assemblages. In highly striated urban and rural landscapes, territories are typically defined by functionalities, and by lines, or boundaries, defined on the earth or inscribed on maps. However, for Deleuze and Guattari, drawing from nomadicism, territory is not necessarily a home, or a stable and bounded space. Striated space, particularly in cities, can be deterritorialized and reterritorialized by lines of flight, or forces that are comparable. In other words, there are latent forces internal and external to any system that can result in dramatic change, in a city these often occur as a result of bureaucratic action or inaction.

Ideally, territories are resistant to precise definition, as territoriality is created through continuous processes of deterritorialization and reterritorialization.[37] As Deleuze and Guattari note, the smooth "always possesses a greater power of deterritorialization than the striated."[38] Augmenting this, DeLanda suggests that any process "which either destabilizes spatial boundaries or increases internal heterogeneity is considered deterritorializing."[39] Therefore, the interrelationship between a territory and the forces of transformation is inherent to the spatial definition of land. This complex concept of a territoriality suggests that striated, or subdivided land, can move towards a smooth space condition under particular circumstances; this refers to the functioning of linear systems, such as boundaries and infrastructure, in a landscape, as smooth space "is a field without conduits or channels."[40] Highly rigid and impenetrable boundaries resist deterritorialization, whereas permeable or shifting boundaries can allow for territorial or spatial reorganization. Deleuze and Guattari refer to the patchwork as an example of a striated system that also can function like a smooth space, patchwork systems are a strong characteristic of agricultural and urban landscapes.[41]

With the extensive use of corridor systems and divisions (property lines, zoning, infrastructure) in urban environments, the result is inordinate lengths of edges. Forman writes: "in sustainability issues, humans are edge species by carving up the land and increasing edges enormously, we eliminate the key values of large patches, thus degrading our landscapes."[42] In other words, when humans settle a landscape they create an infinitude of abrupt and precise edge conditions between landscape elements (blocks, streets, parks, fields, etc.). The porosity of edges between landscape

---

[37] J. MacGregor Wise, "Assemblage," in *Deleuze: Key Concepts*, ed. Charles Stivale (Montreal: McGill-Queen's Press, 2005), p. 79.
[38] Deleuze and Guattari, *A Thousand Plateaus*, p. 480.
[39] M. DeLanda, *A New Philosophy of Society* (London: Continuum, 2006), p. 13.
[40] Deleuze and Guattari, *A Thousand Plateaus*, p. 371.
[41] Ibid., p. 476.
[42] Forman, *Land Mosaics*, p. 81.

elements and the structure of the boundaries or edges (sharp, blurred, or overlapping) are critical to the overall functioning and interconnectedness of a landscape. The boundary between elements determines the amount of flow between adjacent patches of land. The more complicated the boundary, the greater the likelihood of complex interchange.[43] A network can have rhizomatic or arborescent properties depending on its structure, and a network can be either independent of or integrated with surrounding networks and territories. That which flows along a corridor, usually as part of a network, plays a particular function in a larger territorial system. The same can be said for cross flows that encounter boundaries (or corridors). Corridor and boundary systems play a very large role in the ecology of cities.

---

[43] Forman and Godron, *Landscape Ecology*, p. 177.

CHAPTER 4

Agents and Agencies

Cities and agricultural lands are, according to Gilles Deleuze and Félix Guattari, 'striated' landscapes controlled, or highly regulated, by state organizations. Borrowing from landscape ecology,[1] we can provisionally describe contemporary cities as complex patchworks (see previous chapter) comprised of segregated parcels of land divided by infrastructure, property structures, and functionality. Cities, typically, are managed by complex bureaucratic and political structures that provide systems of agency. Agency can destabilize the rigid structures of cities, transforming a city into a smooth space of open flows.

ASSEMBLAGES AND AGENCY

Agency is embedded in Deleuze and Guattari's concept of assemblage, this is evident in the original French term *agencement*. According to Ronald Bogue an assemblage is an arrangement of things and the act of arranging those things also a process of "agencing."[2] Like assemblages, agencies focus on function and action. Conventionally, agency is an intermediate condition, it can be an instrumentality, organization, spatiality, or personification. An agency, which encompasses agents, tends to seek the shortest path through a system in order to effect a change, solution, or project; agents produce effects.[3] However, many other conditions provide faculties or facilities for action, or for the activities of agents. The diagram of an assemblage in effect describes the kind of agency that may be at play, typically based on a desiring function. Desire is, according to Deleuze and Guattari, circulating energy that produces connections, the connections between the components of a given assemblage.[4] The operations of power in systems are driven by desire, Deleuze and Guattari describe individuals as "desiring-machines" that in social groups create "social-machines."[5] The agencing properties of assemblages depend on the ability of a given assemblage to be productive and is a key aspect of an assemblage's diagram. In ecological terms this could mean that human interventions contribute, rather than detract, from environments.

---

[1] See, for example: See Richard T.T. Forman and Michel Godron, *Landscape Ecology* (New York: John Wiley & Sons, 1986); Richard T.T. Forman, *Land Mosaics: The Ecology of Landscapes and Regions* (Cambridge: Cambridge University Press, 1995).
[2] Ronald Bogue, *Deleuze's Way: Essays in Transverse Ethics and Aesthetics* (Aldershot, UK: Ashgate, 2007), p. 145.
[3] See *Shorter Oxford English Dictionary* (Oxford: Oxford University Press, 2007), pp. 41-42.
[4] Reidar Due, *Deleuze* (Cambridge, UK: Polity Press, 2007), p. 95.
[5] See I. Buchanan, "Power, Theory and Praxis," in I. Buchanan and N. Thoburn, eds., *Deleuze and Politics* (Edinburgh: Edinburgh University Press, 2008), p. 17.

Manuel DeLanda uses assemblage theory to examine various types and sizes of social structures from persons to nations. DeLanda, discussing assemblages and social institutions, notes that while the intentional actions of individuals can result in the creation of new institutions, these are as likely to emerge as the "unintended consequence" of intentional action, or as a result of forces at play within assemblages.[6] The conscious actions of specific agents or agencies can lead to change, however, these are dependent on those individuals or organizations that have unique abilities to shape forces in productive ways. Transformations driven by incremental changes over time within a structure or organization, these can be small or significant, are typically based on the inherent activities of an assemblage. However, revolutionary or catastrophic change can occur either from external pressures or events, or because a system "tips" over; a combination of these can occur as a result of shifting alliances between organizations.

As DeLanda also points out, assemblages, as defined by Deleuze and Guattari, are both autonomous entities, and are also related to other assemblages to form complex and shifting alliances. Out of these emerge new organizations, concepts, behaviors, structures, and the like. The agencing properties within a structure, or assemblage, are numerous. Changes within social, political, and physical structures, such as cities, can happen in many ways.[7] There are two categories of agency that will be primarily examined here. First, are individual agents able to seize upon and shape the complex forces at play in an environment and effect change. Second, are the bureaucratic, or organizational, structures that are mandated to design and manage urban environments. There is a third category, that will be touched upon, that involves forces inherent within a system that decode or destabilize a structure, these are the forces of self-organization and deterritorialization that shift boundaries in organizations or on the ground.

When individuals, or singularities, gather into groups they create social machines, as assemblages these can be productive or destructive. The governing of cities and nations involves organizations that have content, expressive regimes, and are necessarily territorial in their functioning. One such example is the state, or the "transcendent imperial machine" which, according to Deleuze and Guattari in *Anti-Oedipus: Capitalism and Schizophrenia*, has a primary role to "overcode."[8] The overcoding by states is a result of highly centered and hierarchical governing systems, leading to the control of territory, the regulation of production, and the striation of space. The role of "decoded flows" is to act as a countervailing force, typically flows that operate outside of established channels, or manipulate weaknesses within

---

[6] See M. DeLanda, *A New Philosophy of Society* (London: Continuum, 2006), p. 24. See also J. Van Wezemael, "The Contribution of Assemblage Theory and Minor Politics for Democratic Network Governance," in *Planning Theory*, 2008, 7:165, pp. 165-185.
[7] DeLanda, *A New Philosophy of Society*, pp. 40-46.
[8] Gilles Deleuze and Félix Guattari, *Anti-Oedipus: Capitalism and Schizophrenia* (Minneapolis: University of Minnesota Press, 1983), p. 199. This terminology is somewhat modified in Deleuze and Guattari's *A Thousand Plateaus*, see E.W. Holland, "Deterritorializing "Deterritorialization": From the "Anti-Oedipus" to "A Thousand Plateaus," *Substance*, Vol. 20, No. 3, Issue 66 (1991), pp. 55-65.

established systems.[9] Deleuze and Guattari identify other social machines at play, including the "territorial machine" which "consists in coding the flows on the full body of the earth," and the "modern immanent machine" which is driven by the decoding flows of capital. The operations of the state and of capital perform deterritorializing functions.[10] Each of these machinic entities has diagrams that describe its operational tendencies. The territorial machine most closely resembles a well-functioning ecology, however, the interventions of state and urban systems operate against this.

Inherent to states are the many bureaucratic systems or agencies that manage everything including land, transportation, banking, education, medicine, etc. These agencies perpetuate the tendencies of the state, with the appearance of being responsive to public need, what were originally decoding flows becomes the new coding system. Bureaucratic and regulatory agencies, are each an assemblage inter-relating with other assemblages, and each has a coded diagram that largely determines how it will act or function. Following Deleuze and Guattari's writings we can briefly describe that capitalism, beyond creating abstract flows of money, production, and consumption, erodes previous forms of control to establish new ones. According to Reidar Due the "profound ambivalence of capitalism and desire makes it difficult to articulate any such space of agency or creation."[11] As Due suggests capitalism is self-serving, it decoded previous forms of state control, resulting in deterritorializations, or boundary shifting, and the creation of new forms of enslavement.[12] Deleuze and Guattari cite the forces of privatization and the creation of abstract wealth as examples of deterritorialization, as capitalism operates based on the abstract flow of capital.

The diagrams that govern the multitude of assemblages that comprise a city tend to mimic the operations of the state, in particular the striation of space; this pertains to both to territories and organizations. As ecologies, cities tend to be highly segregated and inefficient, whereas, smooth space systems tend to be closer to effectively functioning ecologies. To create more ecologically effective cities requires modifying boundary systems within bureaucratic and territorial assemblages, or to move towards smooth space functioning with stronger connectivity and innovation.

INDIVIDUALS VERSUS ORGANIZATIONS

Deleuze and Guattari suggest that certain figures, mainly artists, writers, and philosophers, can act as agents shaping and directing forces, altering boundaries, and causing disturbance. These are individuals operating in fields of forces who look for new diagrams, or new maps of possibilities. Deleuze and Guattari also acknowledge that art, philosophy, and science each have particular modes of operation or agency:

---

[9] Deleuze and Guattari, *Anti-Oedipus*, pp. 222-240.
[10] Ibid., p. 261.
[11] Due, *Deleuze*, p. 114.
[12] See Deleuze and Guattari, *Anti-Oedipus*, pp. 222-224.

art creates affects, philosophy creates concepts, and science creates functions.[13] In their text *Kafka: Toward a Minor Literature* they address the unique work of a Czech Jewish author who wrote in German. Kafka, who himself was a minor functionary, often addressed the seemingly nonsensical nature of bureaucratic and judicial systems. As one would expect, Deleuze and Guattari read Kafka against conventional interpretations. Claire Colebrook notes that their reading of *The Trial*, rather than being understood as an allegory of the law, can be read as the law "exposed as an effect of action, not as some ultimate goal or origin that drives action."[14] In other words, K's labyrinthian movements through the corridors of power creates a new kind of law, or a new form of agency. This is consistent with nomadicism, where movement in space causes constant territorial reorganizations. K operates in a way that exposes the fallacies of a legal system, and yet proposes an alternative course of action; he is a nomadic or destabilizing force. In a similar manner, Deleuze and Guattari describe Henry Miller's traverses through Brooklyn as being like the nomadic reorganization of striated space.[15]

In his text on the British painter Francis Bacon, Deleuze addresses issues of creativity, action, and authorship. Commenting on the work of Bacon, Cezanne, and the abstract expressionists, among others, he writes: "They [painters] say that the painter is already in the canvas, where he or she encounters all the figurative and probabilistic givens that occupy and preoccupy the canvas. An entire battle takes place on the canvas between the painter and these givens."[16] Bacon renders visible a set of forces as they play themselves out against the human figure. In the work of Bacon, which deforms space, objects and figures, the diagram "introduced or distributed formless forces throughout the painting,"[17] it is a "scrambling" function that results in new arrangements and affects. Here, the actions or activities of an individual can under certain conditions provide a new diagram, a new alignment of forces, a new agency; the diagram is successful if something new emerges from it.[18] Forces are at play that are rendered visible or verbal by the artist.

This can be extended to the actions of all individuals that are operating, almost blindly, in a field of forces. However, except under special circumstances, such as the work of a Kafka or a Bacon, the individual as an agent of change is downplayed in the work of Deleuze and Guattari, which is consistent with their ontology of universal becoming and interconnection. Individuals, communities, and organizations have singular qualities, but they are also assemblages, these dovetail with a multitude of other assemblages, each is distinct and yet participates in larger wholes.[19] The individual as an agent of change, can and does occur, but relatively rarely. In theory

---

[13] See Gilles Deleuze and Félix Guattari, *What is Philosophy?* (New York: Columbia University Press, 1994), pp. 201-218.
[14] Claire Colebrook, *Gilles Deleuze* (London: Routledge, 2002), p. 138.
[15] Gilles Deleuze and Félix Guattari, *A Thousand Plateaus: Capitalism and Schizophrenia* (Minneapolis: University of Minnesota Press, 1987), p. 482.
[16] G. Deleuze, *Francis Bacon: The Logic of Sensation* (Minneapolis: University of Minnesota Press, 2005), p. 81.
[17] Ibid., p. 127.
[18] Ibid., p. 128.
[19] DeLanda, *A New Philosophy of Society*, p. 30.

these can be leaders, community activists, and the like, who occupy specific positions and can shape fields of forces or destabilize boundaries that prevent systems from reorganizing.

Cities, as products of the state and central to capitalism, are complex ecologies, as mentioned above. Typically, governmental agencies act to protect an existing system. However, inevitably there are faults in every system, and systems can undergo incremental (evolutionary) change and/or sudden (revolutionary) change. There are forces both internal to and external to a system that can trigger these inversions:

> State societies, far from being the monolithic entities they purport to be, are fissured by nomadic trajectories of various sorts, including those of diverse peripatetic groups the peripatetic populations that roam state societies and escape their regulation are functional components of those societies, not extrinsic exceptions to their control.[20]

Assemblages contain social, expressive, structural, and territorial elements. Socially, these include groups that are resistant to assimilation, or operate against the tendencies of states and capitalism: aboriginals (former nomads), the homeless, gypsies, drop-outs, counter-culture groups, the insane, the impoverished, groups forming any potential countervailing force. For Deleuze and Guattari, an organization or individual tends to lie at the intersection of many forces, and behavior is the result of a complex of inter-relationships. However, they recognize that there are figures, organizations, and forces that do carry the potential to provide a destabilizing or creative force. Generally, Deleuze and Guattari champion the nomadic condition of smooth space, where the nomad continuously makes and remakes boundaries, the nomad is an agent of destabilization.

Like the physical structure of a city, many social and bureaucratic assemblages are clearly bounded. The stability or instability of a territory or organization, depends on a complex set of arrangements. As DeLanda notes, any process "which either destabilizes boundaries or increases internal heterogeneity is considered deterritorializing."[21] He also writes: "One and the same assemblage can have components working to stabilize its identity as well as components forcing it to change or even transforming it into a different assemblage. In fact, one and the same component may participate in both processes by exercising different sets of capacities."[22] The modifying or blurring of boundaries is an act of deterritorialization, it is a agencing that can lead to new arrangements, for example, converting a striated patchwork to a smooth space. Specific kinds of agency, attuned to the complex forces at work in a system, have the potential to make change, such as the ecological transformation of contemporary cities.

---

[20] Bogue, *Deleuze's Way*, p. 117.
[21] DeLanda, *A New Philosophy of Society*, p. 13.
[22] Ibid., p. 12.

ORGANIZATIONS AND BOUNDARIES

Organizations control their boundaries to differing degrees, depending on the function and structure of the organization. When organizations face internal and external challenges or conflicts they can construct boundaries and impose strict rules, particularly on members, or they can expand and loosen boundary structures. Attempts to control dissenting factors can be done with coercion, by appealing to loyalty, or by absorption, co-option, or rejection. [23] Organizations that restrict movement across boundaries tend to follow hierarchical control models, as opposed to those that allow for free movement. Every organization or territory has agents who take on the role of crossing or spanning boundaries. These agents ensure that the entity is interconnected to other entities, this is often necessary for survival. Boundary personnel gather information and materials, and also represent the organization to others. They control or filter the flows in and out of an organization or territory, can create lateral structures, occupy the boundaries of the entity, and store and give off information as required. As "gatekeepers" these personnel or species can hold more power in the overall organization or territory than many of those occupying more central positions; they are also more open to corruption. [24] Therefore, we can suggest that there are various agents involved in the destabilization, transgression, or erasure of boundaries. These include figures who patrol boundaries and yet are impervious to them, and also invasive plant and animal species.

The qualities of boundaries, both in terms of their relationship to entities and as entities themselves, are a vital aspect of the functioning of an organization. Like all spatial conditions, these are in a constant state of change as forces and flows across and along the boundary vary. The edge of a territory or organization, or the space influenced by a boundary, denotes a zone of influence. This is true of both the boundary as a habitat condition, but also for political, geographical, and organizational boundaries. State systems, corporations, and other societal organizations are governed by codes, authority figures, structural dynamics, and jurisdiction, these determine to a large extent the functioning of boundaries. Writing about jurisdiction and boundaries, DeLanda states: "Any process that calls into question the extent of legitimate authority, such as a clash between organizations with overlapping jurisdictions, can destabilize their boundaries, and if the conflict is not resolved, compromise their identity." [25] As DeLanda also points out, conflict can result in the sharpening of boundaries, or differences between insiders and outsiders (eg. citizens versus foreigners). [26] Thus occurs in urban communities, cultural organizations, and religious groups where certain narratives are used to maintain and extend boundaries. Border skirmishes occur when one group wants to test the boundaries of another group. Various "practices of inclusion and exclusion" define

---

[23] H.E. Aldrich, *Organizations and Environments* (Englewood Cliffs: Prentice-Hall Inc., 1979), pp. 244-246.
[24] Ibid., pp. 248-264.
[25] M. DeLanda, *A New Philosophy of Society*, p. 74.
[26] Ibid., p. 58.

the boundaries of groups, which in turn lead to enforcement systems (boundary definition) that include codes, policing systems, territorial marking, and the like.[27]

Organizations and agencies, particularly bureaucratic ones, are continuously being reorganized, usually to address changing circumstances. As well, new agencies emerge, and old ones disappear, or are collapsed into other units; the boundaries of these entities are continuously changing. Like geographical territories, organizational boundaries abut or overlap one another; boundaries can be very fluid and are constantly shifting to adjust for new markets, competition, priorities, etc. Organizations can be subject to the same territorial challenges that operate between adjoining states, jurisdictional ambiguities are a common problem, often resulting in inefficiency and conflict. Organizations are typically charged with functional, and societal, responsibilities, innovation and creativity are not necessarily factors. Agencies are subject to conflicting pressures from all sides, and often struggle to coordinate activities with other organizations; for example, this occurs within the departmental structure of a large corporation or institution.[28] Bureaucracies and administrative systems tend to be clumsy and inefficient, often responding to the force with the most power (often economic). Governmental agencies are often caught between citizens, and large economic forces (corporations); regulatory regimes are susceptible to being outdated and out-moded, unable to address changing circumstances.

The nomadic, or smooth space, definition of boundaries tends to be based on organizational and shamanic systems, rather than constructed and territorial systems. This underscores the emphasis placed on becoming in the work of Deleuze and Guattari. Aldrich defines organizations as "boundary-maintaining systems of human interaction."[29] Organizations are often preoccupied with determining membership, or who belongs and who does not, this is managed by systems of authority and/or governance; this in turn leads to the control of behavior.[30] The management of organizational boundaries, which are often shifting, addresses cross border movements (entry and exit), and interactions with internal and external agencies. As Aldrich notes: "Control over exit is also a defining characteristic of an organization's authority. Authorities have the power to sanction deviants or remove undesirables, and the ultimate sanction they wield is the expulsion from the organization."[31] This suggests that those who are deemed not to belong can be banished, or exiled, or placed in marginal situations such as detention centers. The control of criminal behavior is a major role of the state, much of criminal activity attempts to cross various kinds of boundaries illegally.

---

[27] Ibid., p. 66.
[28] See M. FitzSimmons and R. Gottlieb, "Bounding and Binding Space: The Ambiguous Politics of Nature in Los Angeles," in A.J. Scott and E.W. Soja, eds., *The City: Los Angeles and Urban Theory at the End of the Twentieth Century* (Berkeley: University of California Press, 1996).
[29] Aldrich, *Organizations and Environments*, p. 219.
[30] Ibid., pp. 219-221.
[31] Ibid., p. 224.

Conclusion: Towards Boundaryless Organizations

Generally, Deleuze and Guattari champion the nomadic condition of smooth space, where the nomad continuously makes and remakes boundaries, the nomad is an agent of destabilization. The question arises as to whether agencies and agents that are coded by a particular system, bureaucratic, capitalist, or otherwise, can transform the same system? In the case of contemporary cities, facing a wide range of environmental issues, the re-structuring of boundaries may be possible.

In management theory, the boundaries of organizations are also examined. Some of the most important players in the functioning of organizations (bureaucracies, corporations, etc.) are the boundary personnel (purchasing, marketing, shipping, receiving, etc.), or those who play a vital role in interfacing and providing a linkage to other organizations. Writing about boundary personnel in contemporary business, Aldrich and Herker state:

> Innovation and structural change are often alleged to result from information brought into the organization by boundary personnel. All complex organizations have a tendency to move towards an internal state of compatibility and compromise between units and individuals within the organization, with a resultant isolation from external influences. This trend can jeopardize the effectiveness and perhaps the survival of the organization, unless the organization is effectively linked to the environment through active boundary personnel.[32]

This reflects the inherent tendency in many organizations to look from a center outwards, firstly to the boundaries of the organization and then to external agencies and territorialities. Without, an effective two-way mechanism in play, and an ability to understand the broad landscape, an organization can become isolated and obsolete. The diligent protection of boundary systems normally found in contemporary organizations, provides coherence, but is a movement towards systems with softer boundaries, shifting boundaries, or no boundaries at all, a way to a more effective, and productive, management of territories, cities, bureaucracies, and organizations?

Nick Marshall presents alternative models for organizations and their approaches to boundary management, he opposes these to classical models that support clearly defined boundaries that provided containment. In discussing the notion of boundaries as permeable membranes, he writes:

> Rather than being closed off and self-sufficient entities, organizations are envisaged as open systems which cannot be insulated from the outside world because their boundaries are necessarily and continuously crossed by inputs and outputs, the character of which impose constraints and contingencies relative to the technological and task environments of the organization. In

---

[32] H. Aldrich and D. Herker, "Boundary Spanning Roles and Organization Structure," in *The Academy of Management Review*, Vol.2, No. 2 (April 1977), p. 219.

this sense, the outward-facing boundaries of organization are considered less like the solid walls of a container and more like a permeable membrane or zone of interaction.[33]

Marshall examines socio-cultural construction of boundaries, and the move towards boundaryless, or highly networked, systems. In a world of continuously changing boundaries, he suggests that "alternative and overlapping boundaries are produced, reproduced, enforced, merged, or transcended."[34]

In their analysis of the boundaryless organization, Ashkenas, Ulrich, Jick, and Kerr argue that while you cannot completely eliminate all boundaries in organizations, effective organizations can make vertical, horizontal, external, and geographic boundaries much more fluid and permeable.[35] This is necessary for organizations to have speed, flexibility, integration, and innovation.[36] Post-World War II organizations (and territorial systems) have tended to proliferate division, through the creation of numerous departments, or internal boundaries; this was led by increasing specialization, but could result in problems such as slow organizational operations, protectionism (turf disputes), and lack of integration.[37] However, as Ashkenas, Ulrich, Jick, and Kerr state,

> by making specific external boundaries more permeable, you can dramatically increase speed, flexibility, integration, and innovation. In addition, the more that strategy, technology, management practices, resources, and values flow back and forth naturally, the less necessity there is for crisis-generated breaches of the outer wall. By concentrating on the value chain and the process by which organizations link together to create products and services that have more value combined than separate, you can find a reasonable level of permeability.[38]

In developing organizations that are boundaryless, or have introduced permeability in all boundary situations, it is necessary to work with complexity, ambiguity, continually changing circumstances, and new modes of functioning; older models of corporate control have become obsolete.[39] Successful contemporary organizations are nimble, and able to work with permeable boundaries that are constantly reorganizing, they are not based on top down hierarchical structures with numerous internal divisions. This conception of the boundaryless organization is very reminiscent of the

---

[33] N. Marshall, "Identity and Difference in Complex Projects: Why Boundaries Still Matter in the "Boundaryless" Organization," in N. Paulsen and T. Hernes, eds., *Managing Boundaries in Organizations: Multiple Perspectives* (NewYork: Palgrave MacMillan, 2003), p. 59.
[34] Ibid., p. 71.
[35] R. Ashkenas, D. Ulrich, T. Jick, and S. Kerr, *The Boundaryless Organization: Breaking the Chains of Organizational Structure* (San Francisco: Jolley-Bass, 2002), pp. 2-3.
[36] Ibid., pp. 6-8.
[37] Ibid., pp. 111-112.
[38] Ibid., p. 185.
[39] Ibid., pp. 304-318.

organization of nomadic cultures and animal packs, as conceptualized by Deleuze and Guattari in *A Thousand Plateaus*.

As Hugh Brody notes, for nomadic cultures boundaries are fluid and unstable, as well as porous; boundaries are not static, they belong to operational systems. He writes:

> Being able to move with accuracy on the ground appears to require a parallel freedom of movement of thought—an absence of constraint, a welcoming of many states of mind, from humour to trancing to drunkenness. A fluidity of boundaries, a porousness of divisions, can be seen as useful and normal.[40]

Deleuze and Guattari note that systems oscillate between striated and smooth space,[41] so that it is possible to conclude that it is not possible to achieve a fully boundaryless or bounded condition.

---

[40] H. Brody, *The Other Side of Eden: Hunters, Farmers and the Shaping of the World* (New York: North Point Press, 2000), p. 254.
[41] See Deleuze and Guattari, *A Thousand Plateaus*, pp. 474-500.

Part 2: Early Garden City

Ecologies of the Early Garden City

CHAPTER 5

Garden City Theory

The publication of *To-Morrow; a Peaceful Path to Real Reform* in 1898 by Ebenezer Howard (1850-1928) gave birth to the Garden City movement, rapidly leading to the establishment of the first official Garden City at Letchworth, Hertfordshire, in 1903. Howard was born in London, the son of a shopkeeper, and was a long-term resident of London. After a basic education he trained in clerical work. In 1871 he journeyed to the United States, where he worked in agriculture in Nebraska, and in Chicago where he was again employed in office work. He returned to London in 1876 to take up employment as a parliamentary reporter, a position he held for the remainder of his working life.

On his return to London, Howard found a city that was undergoing great stress, a city awash in a wide variety of reformers dedicated to changing society.[1] He read widely and by the late 1880s was examining the land question. This refers to the ownership and control of land, an issue that preoccupied many late-Victorian activists, in part as a response to the ongoing crisis in British agriculture.[2] It was also during this period that a number of key intellectuals, many of whom Howard would cite in his book, were developing social theories that united economics, psychology, and political theory, in response to "a second industrial revolution, social dislocation, agricultural decline, unemployment, and mounting discontent [which] exerted a persistent and accelerating pressure upon traditional social theory."[3] The late Victorian and Edwardian period was also one of political change with the emergence of socialism and the British Labour party.[4] The Garden City concept emerged at a time of crisis in both the English city and countryside.[5]

INFLUENCES ON THE GARDEN CITY

In *To-Morrow*, Howard responded to the complex forces at play, he candidly acknowledged the sources for many of his ideas, suggesting that it was a "combination of three distinct projects": 1) to organize the "migratory movement of

---

[1] P. Hall, D. Hardy, and C. Ward, "Commentator's Introduction," in E. Howard, *To-Morrow: A Peaceful Path to Real Reform* (London: Routledge, 2003), pp. 3-5.
[2] See Ibid., pp. 2-3. See also P. Horn, *The Changing Countryside in Victorian and Edwardian England and Wales* (London: The Athlone Press, 1984); J. Marsh, *Back to the Land: The Pastoral Impulse in England, from 1880 to 1914* (London: Quartet Books, 1982).
[3] R.N. Soffer, "The Revolution in English Social Thought, 1880-1914," in *The American Historical Review*, Vol. 75, No. 7 (Dec 1970), p. 1938.
[4] See D. Read, *Edwardian England 1901-15: Society and Politics* (London: Harrap, 1972).
[5] D. Fraser, "The Edwardian City," in D. Read, ed., *Edwardian England* (London: Croom Helm, 1982), p. 60.

population"; 2) to promote a new "system of land tenure"; and 3) to build upon various conceptual models for cities.[6] The notion of the organized relocation of populations from crowded industrial cities into Garden Cities was a direct response to some of the most egregious aspects of the Industrial Revolution, including the creation of urban working-class slums and highly exploitive working conditions. In developing his ideas Howard was inspired by a number of contemporary economists, philosophers, and political theorists including Thomas Spence, Alfred Marshall, Herbert Spencer, Henry George, and Peter Kropotkin. There were also a series of urban models, such as that described by Edward Bellamy in his novel *Looking Backward*, and those proposed by Benjamin Richardson and James Silk Buckingham, that contributed to his vision. Further, examples of communities built at Bournville and Port Sunlight in the late nineteenth century would factor into his conceptual design. He was also aware of the Arts and Crafts Movement as defined by John Ruskin and William Morris, which had such enormous impact on late nineteenth and early twentieth century design practices, however, Howard was never an active proponent of its ideals. From these numerous sources, Howard gradually developed his Garden City model that attempted to unite town and country, address housing and general labor issues, and confront questions of land ownership.[7]

The latter part of the nineteenth century coincided with a depression, particularly in farming and agricultural land values. The Land Nationalization Society emerged in 1881 in response to a range of land ownership issues, the movement promoted the concept that land should be held in common, and not in private hands, and that the countryside should be repopulated. The land question was exacerbated by a shortage of good housing, particularly in London where over-crowding occurred; this was a result of the steady depopulation of the countryside that had occurred since the Industrial Revolution as the rural poor sought work in the cities. The work of the prominent British economist Alfred Marshall (1842-1924) greatly influenced Howard's ideas in this area, although Marshall would subsequently not support the land nationalization movement. In particular, Marshall's 1884 pamphlet "Where to House the London Poor?" which advocated for the coordinated movement of population from London to the country was important for Howard. In the text Marshall outlined the disadvantages of maintaining manufacturing industries and workers in large cities, such as London, when there are the means for relocating these to the country, and "combining the advantages of the town and the country."[8] Describing the declining benefits of locating industry in over-crowded cities, Marshall wrote:

> But as the century [nineteenth] wore on, and communication was opened up, the special advantages which residence in large towns offered to producers gradually diminished. Railways, the cheap post, the telegraph, general

---

[6] Howard, *To-Morrow*, p. 103.
[7] See M. Miller, *Letchworth: The First Garden City* (Chichester: Phillimore & Co. Ltd., 1989), pp. 10-14.
[8] Reprinted as A. Marshall, *Where to House the London Poor* (Cambridge, W. Metcalfe and Son, 1885), p. 4.

newspapers and trade newspapers, and organized associations among employers and employed, all had a share in the change. Meanwhile space in the towns was becoming more and more valuable for trading and administrative and other purposes; and manufacturers began to doubt whether the special advantages of the town were worth the high ground-rents that they had to pay there.[9]

Marshall suggested that it could be "economically advantageous" to relocate industries to the country by using modern technology, and alleviate urban overcrowding, although he did acknowledge that certain manufacturing was better suited to large cities. In his textbook *Principles of Economics*, first published in 1890, Marshall addressed a wide range of topics including agents of production (land, labor, capital, and organization), the fertility of land, population growth, population health, and the rent of land; all of these factored directly into Howard's development of the Garden City model. It is likely that Howard was familiar with this text and was acquainted with Marshall personally.[10]

Howard was involved in the land nationalization movement and was a keen proponent of land reform. Howard's original source of inspiration for this was an obscure pamphlet entitled *The Rights of Man, as Exhibited in a Lecture*, read at the Philosophical Society in Newcastle, published in 1775 by a radical named Thomas Spence (1750-1814). The historian T.M. Parssinen writes that, according to Spence:

> At some indeterminate time in the past, a few men simply claimed the land, and those who were victimized never questioned these claims. Control of the land meant control of the lives of the men who depended on it. Dominion over the land was extended to dominion over man; the landlords became tyrants.[11]

Based on the territorial concept of the church parish, Spence proposed that citizens would organize themselves into corporations, which would then be the sole owner of land, all rents would accrue to the corporation, there being no taxes or duties.[12] This model of a common ownership of land and generating public revenues from rent was very similar to that adopted by Howard for the Garden City.

Also arguing against the concept of the private ownership of land, and an influence on Howard, was the prominent Victorian philosopher and political theorist Herbert Spencer (1820-1903). In his book *Social Statics*, first published in 1851, he championed personal liberty and freedom, and examined concepts of happiness and morality. As part of his overall thesis, he wrote:

---

[9] Ibid., pp. 3-4.
[10] Hall, Hardy, and Ward, "Commentary," in Howard, *To-Morrow*, p. 23.
[11] T.M. Parssinen, "Thomas Spence and the Origins of English Land Nationalization," *Journal of the History of Ideas*, Vol. 34, No. 1 (Jan.-March, 1973), p. 136.
[12] Ibid., p. 136.

> Not only have present land tenures an indefensible origin, but it is impossible to discover any mode in which land can become private property. Cultivation is commonly considered to give a legitimate title. He who has reclaimed a tract of ground from its primitive wildness, is supposed to have thereby made it his own. But if his right is disputed, by what system of logic can he vindicate it?[13]

Spencer argued that labor does not entitle one to the ownership of land.[14] He suggested that "all men have equal rights to the use of the earth."[15] He acknowledged the logistical challenges of this notion, but upheld the idea that "every man has freedom to do all that he wills, provided he infringes not the equal freedom of any other man."[16] His ideas extended to the constitution of the state, and to the role of governments.

Another influential advocate for land nationalization was the American social economist and philosopher Henry George (1839-1897). In his popular book *Progress and Poverty*, published in 1878, George attempted to address the "cause which produces poverty in the midst of advancing wealth."[17] He began his text by asking how the nineteenth century, a period that produced such wealth, had also created such poverty and idleness?[18] He noted that the wealth produced by the industrial revolution had driven a wedge between the rich and the poor.[19] George argued that wealth is divided between wages, rent, and interest.

George identified several factors contributing to the concept of material progress. The first addressed the effect of increased population, which would result in increased rents and lowered wages.[20] Secondly, when it comes to "improvements in the arts," or human invention, the development of labor saving devices, allows labor to produce more wealth, increasing the demand for land and rent.[21] George argued that the increased production of wealth, leading to increased land rents, and as a result, land speculation. He charged that land speculation is a major cause of economic depressions, particularly land speculation at the edge of cities:

> when we reach the limits of the growing city we shall not find the land purchasable at its value for agricultural purposes we shall find that for a long

---

[13] H. Spencer, *Social Statics, or the Conditions Essential to Human Happiness Specified and the First of The Developed* (New York: Augustus M. Kelly Publishers, 1969), p. 116.
[14] Ibid., p. 127.
[15] Ibid., p. 130.
[16] Ibid., p. 217.
[17] H. George, *Progress and Poverty: An Inquiry into the Cause of Industrial Depressions and of Increase of Want with Increase of Wealth, The Remedy* (New York: The Modern Library, 1938), p. 17. See also, S. Buder, *Visionaries and Planners: The Garden City Movement and the Modern Community* (New York: Oxford University Press, 1990), for a discussion of George's influence on the Garden City movement.
[18] George, *Progress and Poverty*, p. 6.
[19] Ibid., p. 9.
[20] Ibid., p. 234.
[21] Ibid., p. 249.

distance beyond the city, land bears a speculative value, based upon the belief that it will be required in the future for urban purposes.[22]

He argued that this speculation in the value of land (rent) drove down wages and interest, leading to reductions in production.[23] After his lengthy analysis, George's "true remedy" for the "unequal distribution of wealth" was "to substitute for the individual ownership of land a common ownership."[24] The role of governments would be to administer infrastructure and common property for "a great cooperative society."[25] George's writings, along with those of Marshall, Spence, and Spencer, would directly influence some of Howard's important ideas for the Garden City, such as the common ownership of land through the creation of a company. However, there were other important thinkers who addressed related issues of labor and production.

In the late 1880s it is likely that Howard read essays by Peter Kropotkin (1842-1921), the influential anarcho-communist, although he is not cited in Howard's book.[26] Kropotkin, who lived a rather tumultuous life, resided for many years in England, beginning in 1886. An advocate for a free society without a state system, his writings bear some similarity to Howard's vision. Kropotkin's early essays, which were subsequently collected in the book *Fields, Factories and Workshops*, published in 1899, added up to a sustained study of industrial and agricultural production that examined the "division of labor," and the centralization of industry. Kropotkin described the concept of uniting both urban and rural forms of meaningful labor in the following terms:

> We proclaim integration; and we maintain that the ideal of society—that is, the state towards which society is already marching—is a society of integrated, combined labour. A society where each individual is a producer of both manual and intellectual work; where each able-bodied human being is a worker, and where each worker works both in the field and industrial workshop; where every aggregation of individuals...produces and itself consumes most of its own agricultural and manufactured produce.[27]

The idea of meaningful labor is central to the Garden City model, particularly in attracting enlightened employers and integrating the act of gardening. In the essays Kropotkin undertook a detailed study of European modes of production, both industrial and agricultural, he noted that while British agriculture had been innovative and highly productive, by the 1890s it was in decline, with fewer persons per capita

---

[22] Ibid., p. 257.
[23] Ibid., p. 264.
[24] Ibid., p. 328.
[25] Ibid., p. 456.
[26] See Hall, Hardy, and Ward, "Commentator's Introduction," in Howard, *To-Morrow*, p. 4.
[27] P. Kropotkin, *Fields, Factories and Workshops, or, Industry Combined with Agriculture and Brain Work with Manual Work* (London: Thomas Nelson & Sons, 1913), p. 23.

employed in agriculture than continental Europe. This was a result of factors such as concentrated land ownership, an emphasis on profit, and a lack of cooperation.[28]

Kropotkin also described a great increase in market gardening in Europe, and Great Britain, during the late part of the nineteenth century and into the beginning of the twentieth century. This movement towards agriculture produced from small holdings was consistent with Howard's Garden City ideas. Kropotkin wrote:

> It has been proved that by following the methods of intensive market-gardening—partly under glass—vegetables and fruit can be grown in such quantities that men can be provided with a rich vegetable food and a profusion of fruit, if they simply devoted to the task of growing them the hours which everyone willingly devotes to work in the open air, after having spent most of his day in the factory, the mine, or the study.[29]

Kropotkin argued that in older communities agricultural and industrial labor had always been united. In many European countries, workers cultivated small agricultural holdings, and during the winter months were also engaged in small industrial activities, or trades (undertaken at home, or in small workshops).[30] Citing models developed in Russia, he advocated for integrated education that united manual and "brain" labor, based on scientific and technical learning. In the second edition of Kropotkin's book, published in 1913, he acknowledged the Garden City as a model of integrated labor.

With respect to visionary models for new towns Howard was influenced by writers such as Benjamin Richardson (1828-1896), Edward Bellamy (185-1898), and James Silk Buckingham (1786-1855). Richardson's scheme for the healthy city of Hygeia is referenced twice in Howard's text. Richardson described his plan in an address he delivered in 1875 in which he provided many practical examples for creating healthy and sanitary living environments.[31] One of the greatest sources of inspiration for Howard was Bellamy's immensely popular novel *Looking Backward*, published in 1888, Howard was so enamored with the book that he encouraged a London publisher to produce an edition. The novel follows a young Bostonian named Julian West who traveled in time from 1887, to the year 2000. Awaking from a long sleep, West discovers an egalitarian and cooperative society managed by a benevolent state, and a city defined by greenery and contentment, West describes the city as:

> Miles of broad streets, shaded by trees and lined with fine buildings, for the most part not in continuous blocks but set in larger or smaller enclosures, stretched in every direction. Every quarter contained large open squares filled with trees, along which statues glistened and fountains flashed in the late-afternoon sun. Public buildings of a colossal size and architectural

---

[28] Ibid., p. 97.
[29] Ibid., p. 411.
[30] Ibid., pp. 241-245.
[31] See B.W. Richardson, *Hygeia: or a City of Health* (London, MacMillan and Co. Ltd., 1876).

grandeur unparalleled in my day raised their stately piles on every side. Surely, I had never seen this city nor one comparable to it before.[32]

West is guided by Dr. Leete, his host, and his daughter Edith, who provide detailed instruction in the ways of a world unfamiliar to him. The text provided a vision of a society that had seemingly solved the many problems that plagued the nineteenth century. In the end, West awakens from a dream back in the crowded and intense environment of nineteenth century Boston. The Boston of 2000, described in the text, is a Gilded Age city of greenery, fine architecture, and efficiency. Bellamy's book provided Howard with a vision of an egalitarian society based on co-operation rather than self-interest, although it lacked many of the pragmatic details for establishing a new community that Howard was so devoted to.[33]

In his publication *National Evils and Practical Remedies, with the Plan of a Model Town*, published in 1849, James Silk Buckingham, an English parliamentarian and writer, also provided the description for a model town. The plan shows a one-mile square town organized around a central square which supports various administrative institutions, this is surrounded by a layer of cultural institutions, and streets of housing. The plan, which is similar in formal organization to the one that Howard proposed, is divided into neighborhoods by radiating boulevards. Beyond the town is 10,000 acres area of farmland. Buckingham described the purpose of the new town in the following terms:

> to combine within itself every advantage of beauty, security, healthfulness, and convenience, that the latest discoveries in architecture and science can confer upon it; and which should, at the same time, be peopled by an adequate number of inhabitants, with such due proportions between the agricultural and manufacturing classes, and between the possessors of capital, skill, and labour, as to produce, by the new combinations and discipline under which its code of rules and regulations might place the whole body, the highest degree of abundance in every necessary of life, and many luxuries, united with the lightest amount of labour and care, and the highest degree of health, contentment, morality, and enjoyment, yet seen in any existing community.[34]

Named Victoria, the town was to include the latest in technology, be owned by a company, be religiously free, and employ enlightened labor practices. C.B. Purdom, who documented the early development of Letchworth, noted the many similarities between Buckingham's and Howard's schemes.[35] Beyond theoretical sources,

---

[32] E. Bellamy, *Looking Backward, 2000-1887* (Harmondsworth: Penguin Books Ltd., 1982), p. 55.
[33] See D. MacFaydyen, *Sir Ebenezer Howard and the Town Planning Movement* (Cambridge, Mass.: MIT Press, 1970), p. 21.
[34] J.S. Buckingham, *National Evils and Practical Remedies, with the Plan of a Model Town* (London: Peter Jackson, Late Fisher, Son, & Co., 1849), p. 141.
[35] C.B. Purdom, *The Garden City: A Study in the Development of a Modern Town* (London: J.M. Dent & Sons Ltd., 1913), p. 22.

Howard also derived inspiration from various nineteenth century model communities, including Port Sunlight built by the Lever Bros. soap manufacturer beginning in 1888 and Bournville built for the Cadbury chocolate company starting in 1893.

As Howard readily acknowledged, he synthesized ideas from a wide range of contemporary sources in order to develop his own vision of the Garden City. Many of these were prominent nineteenth century thinkers involved in economic, social, and political reform. As Soffer argues the period from 1880 to the outbreak of war in 1914, during which the Garden City was both formulated and developed, represents a revolutionary period in English social thought, a period devoted to the development of society scientifically.[36] Beyond these thinkers were those who had developed both theoretical and practical versions of model communities. Howard, in his rationally written text produced a theoretical vision of a community, based on the common ownership of land and cooperation, that would widespread and enduring influence.

EBENEZER HOWARD AND *TO-MORROW: A PEACEFUL PATH TO REAL REFORM*

In Howard's original publication, much of the writing was devoted to the management of the town, and the concept that land would be held in common ownership through a company, and then would be leased by individual tenants. Along with the text, there are a number of now famous diagrams that attempted to define the inter-relationship between town and country, the relationship between a Garden City and a Central City, the organization of the town and surrounding lands, and the basic layout of streets and functions. The following section provides an overview of the contents of Howard's book.

In the introductory section, citing a number of the authors noted above, he raised the issue of over-crowding in London, and questions of urban squalor, rural depopulation, and the "congestion of labour" in cities.[37] Ameliorating labor conditions in both the city and the country was a major concern for Howard and the Garden City movement. In this section he introduced the "The Three Magnets" diagram, which has since become one of the most recognizable urban diagrams in urban design history. Howard described the city as a magnet, as an attractive force drawing people off the land. He asked how this trend could be reversed? His answer lay in neither the country nor the city, but in a union of country and city. Howard, and his followers, clearly intended to draw the labor force away from urban centers and to "restore the people to the land."[38] However, he also wanted to include poor rural laborers, an overlooked segment of the population, into the mix. Challenging what he saw as the unhealthy aspects of dense cities, and proposing alternatives to issues of rent, wages, and the like, he wrote:

---

[36] See, Soffer, "The Revolution in English Social Thought, 1880-1914."
[37] Howard, "Introduction," in To-Morrow, pp. 4-5.
[38] Ibid., p. 5.

> The town is the symbol of society and the country! The country is the symbol of God's love and care of man. All that we are and all that we have comes from it. Town and country must be married, and out of this joyous union will spring a new hope, a new life, a new civilization.[39]

In the first chapter of the book, Howard provided diagrams that show a circular city organized around a central park, and surrounded by a green belt supporting various institutions, forests, and agricultural uses. He described the economic strategy for the development of a Garden City following the purchase of suitable land. The Garden City would be financed through escalating land rents that would go to a Central Council. A 6,000-acre parcel would be acquired, with 1,000 acres devoted to the town, the remainder devoted to the agricultural belt. The town was designed to accommodate 30,000 residents, with a further 2,000 residing in the agricultural belt; the town and surrounding agricultural belt were conceived of as a single entity. The overall plan is formal, concentric, and hierarchical, clearly organized by function, and bearing a resemblance to Buckingham's plan. The centralized organization of the scheme recalls other such plans from urban history, including various utopian city plans from the Renaissance. However, instead of a public square, the center is occupied by a public garden, surrounded by public institutions, and a Central park. This is then encircled by a "Crystal Palace," or covered arcade. The next layer is occupied by housing, and then a "Grand Avenue" which is a prominent green space containing schools and religious buildings. Beyond this lies further housing, an outer ring of industrial functions, and finally the agricultural estate (or belt). The town is sub-divided by a hierarchy of tree-lined boulevards, avenues, and roads into six "wards" or neighborhoods; the idea of the ward as an integrated subset of the town was one of the innovations of the plan. The diagrams show areas allocated to housing but provides no indication of plots or housing arrangements. Howard mentioned that some houses will have gardens, however, there is no detailed description of the role of gardens in the design, nor is the "agricultural belt" described in any functional detail.

The second chapter is devoted to the "agricultural estate," or Howard's theories of land rent, as derived from Spencer and George. In the chapter Howard underscored that a significant difference between the Garden City and existing cities is that:

> Its entire revenue is derived from rents; and one of the purposes of this work is to show that the rents which may reasonably be expected from the various tenants on the estate will be amply sufficient, if paid into the coffers of the Garden City, (a) to pay the interest on the money with which the estate is purchased, (b) to provide a sinking-fund for the purpose of paying off the principal, (c) to construct and maintain all such works as are usually constructed and maintained by municipal and other local authorities out of rates compulsorily levied, and (d) (after redemption of debentures) to provide

---

[39] Howard, *To-Morrow*, p. 48.

a large surplus for other purposes, such as old-age pensions or insurance against accident and sickness.[40]

The central idea of the Garden City, taken from Spence, was that "the city creates its own land values."[41] Howard, reflecting his involvement in the land issue, argued that urbanization dramatically raises land values, and that this value should accrue to the municipality, rather than to speculators. The rent paid by residents of the town will directly benefit them through improvements, and through a local social welfare system; Howard indicated this in his fourth diagram entitled "The Vanishing Point of Landlord's Rent." Turning to the agricultural estate, Howard noted that the development of the Garden City would also raise the value of adjacent agricultural land, and that the local farms will have a secure local market for the selling of produce. This was proposed against the influx of cheap foreign produce that flooded Britain in the late nineteenth century. Further, waste from the town could be used to fertilize surrounding farms. The agricultural portion of the estate would have higher rents than those for surrounding farms, as a tenant of the Garden City agricultural belt would have a planned sewage system, a controlled market, and good transport opportunities to deliver produce to markets.[42] Howard believed that farmers would be attracted to the Garden City, enjoying the natural and healthful qualities of the arrangement.[43]

In the third chapter Howard examined the revenues that could be generated from rents in the town portion of the estate. At the beginning of the chapter Howard quoted from Alfred Marshall's pamphlet "The Housing of the London Poor," regarding the concept of relocating people from overcrowded London to the country. Howard further developed his calculations on rent, and noted that the average house lot would be 20 feet by 130 feet, accommodating on average 5.5 persons per lot, he wrote:

> It [the plan] obtains ample space for roads, some of which are of truly magnificent proportions, so wide and spacious that sunlight and air may freely circulate, and in which trees, shrubs, and grass give to the town a semi-rural appearance.[44]

It is clear from the diagrams and this description that Howard was influenced by various formal models, but that there is also a distinctive American quality, evident in the spaciousness of the scheme, his ideas no doubt inspired by his experiences in the United States.[45] The diagrams describe a town that despite its small size is quite urbane in its organization.

Chapter four begins with a quote from November 1890 issue of *The Echo* that advocated for school gardens so that children could have instruction in horticulture, a

---

[40] Ibid., pp. 20-21.
[41] See Hall, Hardy, and Ward, "Commentary," in Howard, *To-Morrow*, p. 43.
[42] Howard, *To-Morrow*, pp. 26-27.
[43] Ibid., p. 29.
[44] Ibid., p. 32.
[45] See footnote in Ibid., pp. 45-46.

notion that would be developed by the Garden City movement.[46] Attempting to provide a pragmatic basis for his vision, Howard compared the cost of development in existing areas of London, against the costs associated with a new development on inexpensive agricultural land, which would allow for less expensive and better quality housing. He stressed the point that the Garden City would be a modern planned community, involving a wide range of experts.[47] Howard noted that each of the six wards, or neighborhoods, would be largely self-sufficient, and would be built in sequence.[48] Further, a new town could accommodate a full range of underground services for sewage, water, gas, electricity, telephone, etc. All of this would result in significant cost savings, according to Howard. In the fifth chapter Howard further developed the topic of expenditures, by focusing on the role of infrastructure, including roads, schools, institutional buildings, parks and "road ornamentation," and sewage.

The sixth chapter is devoted to the administration of the Garden City. Appealing to a wide political spectrum Howard attempted to strike a balance between the interests of the community and the interests of individuals. For the administration of the Garden City he proposed a Board of Management consisting of an elected Central Council and various Departments. He suggested that the Central Council would have more power than similar municipal authorities, and that it would oversee departments in three areas: public control (finance, assessment, law, and inspection), engineering (roads, parks, drainage, public buildings, etc.), and social purposes (education, libraries, baths and wash-houses, music, and recreation).[49] The administrative model he proposes is centralized, hierarchical, and departmentalized, with clearly defined entities and responsibilities; it incorporates control, technical, and social mechanisms.

Chapter seven addressed a variety of topics, including the notion of "semi-municipal enterprise" which involved private enterprise conducted in publically funded structures such as markets; for example, the Crystal Palace proposed by Howard provided this function. He examined questions pertaining to manufacturers and retail tenants and the "local option," and how businesses may be adequately protected, providing regulated competition, or managing relations between producers and consumers.[50] Finally, in this section Howard addressed the temperance question, which was an important one at the time. Although he acknowledged that the citizens of a Garden City might reject the notion of selling alcohol, Howard suggested that a controlled number of drinking establishments would be his preference, and that this would generate additional revenues. Despite his position, Letchworth would originally be a dry community.

In chapter eight Howard explained the concept of "pro-municipal work," or public service. In one of the most provocative statements in the text Howard discussed the role of public service and the kind of people that would be attracted to the Garden City:

---

[46] Ibid., p. 36.
[47] Ibid., p. 45.
[48] Ibid., pp. 38-39.
[49] Ibid., pp. 67-68.
[50] Ibid., pp. 73-79.

those who have the welfare of society at heart will, in the free air of society, be always able to experiment on their own responsibility, and thus quicken the public conscience and enlarge the public understanding.

The whole of the experiment which this book describes is indeed of this character. It represents pioneer work, which will be carried out by those who have not a merely pious opinion, but an effective belief in the economic, sanitary, and social advantages of common ownership of land, and are impelled to share their views shape and form as soon as they can see their way to join a sufficient number of kindred spirits. [51]

Other examples of pro-municipal work include public housing, building societies, trade unions, and the like.[52] This occurred at a time in Britain where there numerous cooperative, reformist, charitable, and philanthropic organizations at work. The emphasis on a pioneering spirit, suggests that the Garden City, with its relatively low-density fabric, was always intended to be a place where everyone would participate in community governance and culture, that in a "progressive community" local organizations would provide a high level of engagement, as opposed to the way a citizen would participate in a large city; pioneering is equated with public service. Howard acknowledged that a wide range of organizations (religious, political, cultural, philanthropic, and charitable) are necessary in a community, including those societies involved in the financing of housing.

In chapter nine Howard synthesized the three previous chapters with reference to Diagram No. 5 entitled "Diagram of Administration" (subsequently not included in the book republished in 1902 as *Garden Cities of To-Morrow*). Here the Central Council is encircled by the three Departments, more remote are the semi-municipal group (various markets), the pro-municipal group (societies, educational institutions, hospitals, banks, etc.), and then the cooperative and individualistic group (allotments, factories, clubs, farms, etc.). This diagram described the operational system designed to administer the Garden City and is clearly hierarchical in its design. In chapter ten there are "some difficulties considered," and Howard raised the question of previous "social experiments" in developing new communities, many of which failed. Further, he discussed the strengths and weaknesses of Communism and Socialism. He acknowledged that Communism (and Socialism) has strong and important principles, but that a society must be balanced by the recognition of the role of the individual. He concluded the section by arguing that the Garden City takes a pragmatic position and incorporates a wide range of beliefs and enterprises.

Throughout his text Howard drew support from the various figures that had inspired his vision. For example, in the eleventh chapter he quoted Spencer, from his book *Social Statics*, on the concept of common land ownership, in which Spencer wrote:

---

[51] Ibid., p. 83.
[52] Ibid., pp. 84-85.

Instead of being in the possession of individuals, the country would be held by the great corporate body—society. Instead of leasing his acres from an isolated proprietor, the farmer would lease from the nation. Instead of paying his rent to the agent of Sir John and His Grace, he would pay it to an agent or deputy agent of the community. Stewards would be public officials instead of private ones, and tenancy the only land tenure.[53]

While Spencer eventually concluded that his ideas about rent were not feasible, Howard advocated for common land ownership at the local, rather than national, level. Further, Howard argued that, while there were formal similarities between his Garden City model and the plan for a community put forward by Buckingham, the two schemes were organizationally and socially very different. Howard concluded the chapter by restating his objectives, stating that there needed to be an organized migration from over-crowded cities to the country to take advantage of its benefits ("fresh air, sunlight, breathing room and playing room"), that this migration would benefit from common land ownership.[54]

Howard, in the twelfth chapter, imagined that the Garden City was already built, and wished to "consider briefly some of the important effects" of the "pathway to reform."[55] Here there was also a brief mention of the garden, he wrote that society "should forthwith gird themselves to the task of building clusters of beautiful home-towns, each zoned by gardens."[56] Throughout his text Howard attempted to present a vision that could be realized, and to balance collective and individual desires, suggesting that reform should begin with a small scale experiment rather than wholesale change. This foreshadows the development of the first Garden City at Letchworth in 1903, which despite its shortcomings, would be a demonstration of many of Howard's theories; the direct and rational nature of Howard's book would mean that it would become a manual for establishing a model community. His effort to balance Individualism and Socialism was also a key aspect of the vision. He describes Individualism as "a state in which there is a fuller and freer opportunity for its members to do and to produce what they will, and free associations, of the most varied kinds" and Socialism as "a condition of life in which the well-being of the community is safe-guarded, and in which the collective spirit is manifested by a wide extension of the municipal effort."[57] In this chapter Howard cited other writers who shared aspects of his vision, including John Stuart Mill, Henry M. Hyndman, and 'Nunquam' (Robert Blatchford). He also remarked on the enormous technological and societal change that had been effected in society during the previous sixty years. This chapter is philosophical in tone, and in it Howard raised the question of the earth and its "infinite treasures;" he wrote:

---

[53] Herbert Spencer quoted in Ibid., p. 109.
[54] Ibid., p. 114.
[55] Ibid., p. 116.
[56] Ibid., p. 116.
[57] Ibid., pp. 118-119.

> Now, as every form of wealth must rest on the earth as its foundation and must be built up out of the constituents found at or near its surface, it follows (because foundations are ever of primary importance) that the reformer should first consider how best the earth may be used in the service of man.[58]

In the thirteenth chapter provocatively titled "Social Cities" it is evident that Howard considered the Garden City to be an important invention, one which would catch on like other significant inventions in the nineteenth century such as the railway, however, parliamentary intervention would be necessary. His statements foreshadowed the British New Towns Act of 1946, which the Garden City movement directly influenced. The notion of a program of new towns is further reflected in Diagram No. 7, which described a cluster of Garden Cities, separated by green space, around a larger central city (with a population of 58,000). He noted, "any well-planned cluster of towns, must be carefully designed in relation to the site it is to occupy."[59] He clearly understood that his diagrams were illustrations of concepts. Howard argued that through modern transportation citizens would have easy access to the country "higher forms of corporate life," or the benefits of the town. Howard stressed the necessity to build healthy cities "in which there may be ample space and ventilation, and in which modern scientific methods and the aims of social reformers may have the fullest scope in which to express themselves."[60] Expecting to draw upon the talents of a wide range of experts, Howard concluded the chapter by reiterating the benefits of the Garden City with respect to transportation, land tenure, pensions, hope, peace, and goodwill.[61] The fourteenth chapter addressed "The Future of London" and the problems associated with the city at the end of nineteenth century. The fifteenth, and final, chapter was devoted to water supply, in which Howard attempted to demonstrate a more effective system, including the prevention of waste. Dependent on good local supply, the proposal involved a system of reservoirs, canals, and pumping mechanisms (including windmills). He suggested the separation of potable and non-potable water, and examined issues of drainage, irrigation, transportation, power, recreation, and ornament.

Ironically, in the book Howard made very little reference to the idea of the garden or the agricultural belt, devoting most of his efforts to economic and administrative issues.[62] The Garden City movement that would emerge attempted to synthesize what it determined to be the best of town and country. Ultimately, the Garden City would include both aesthetic and reformist ideals, providing economic, social, and cultural opportunities.

---

[58] Ibid., p. 125.
[59] Ibid., pp. 130-131.
[60] Ibid., p. 134.
[61] Ibid., pp. 140-141.
[62] The origins of the term "Garden City" in Howard's writings remains ambiguous, see M. Elen Deming, "The Place of the Garden in Garden Cities: A Utopian Romance," *CELA Proceedings* (November 1996), pp. 38-45.

CONCLUSION: A RURAL-URBAN VISION

According to Howard the nineteenth century town provided places of work, and also required high rents. The industrial town was oppressive with unemployment, slums, and high prices; environmentally it was polluted and removed from nature; the town also created the isolation of the crowd. On the other hand, the country provided the beauty of nature, and sunshine, fresh air and water, but there is little social opportunity in the isolation of the countryside. The Garden City attempted to blend the good qualities of country and town, largely rejecting monumental urbanity in favor of a small rural town/village atmosphere dominated by greenery. In fact, as a satellite to a larger city, the Garden City was, in part, intended to provide a familiar environment to those migrating to the city from the country. The extensive discussion of finances in Howard's text seemed to reflect a desire to appeal to future investors. This is borne out by the fact that the first Garden City at Letchworth came into existence so soon after the publication of the book. There is no doubt, that while the Garden City movement spawned by Ebenezer Howard only realized two garden cities, Letchworth Garden City and Welwyn Garden City, it had widespread influence on subsequent urbanism throughout the twentieth century; within Howard's book, and text and the diagrams provided vehicles for the creation of productive communities.

Ecologies of the Early Garden City

CHAPTER 6

Letchworth Garden City, 1903-1913

The emergence of the Garden City movement, inspired by Ebenezer Howard's book *To-Morrow: a Peaceful Path to Real Reform* (1898), subsequently re-published as *Garden Cities of To-Morrow* (1902), would have an enormous impact on future urban development and town planning worldwide.[1] Lewis Mumford claimed that the two most important inventions of the early twentieth century were the airplane and the Garden City.[2] The Garden City model in many ways represents the anti-thesis to the historic city, as a model derived from smaller rural communities, with a defined size, low densities, and a wealth of green space. Many subsequent urban models have expanded upon, altered, and diverged from Howard's ideas. The Garden City radically challenged the expectation that a city is a dense, vibrant, and largely hard landscaped environment. In fact, urban environments developed over the last half century have in many cases been dispersed, low intensity, and soft landscaped environments, resulting in substantial changes to the way cities are constructed, managed, and inhabited. No doubt, the impact of the Garden City movement remains controversial, and is blamed for the rise of suburbia and for pitting design professionals against communities and land developers.

THE EARLY DEVELOPMENT OF LETCHWORTH

The first decade of Letchworth's history provides a case study for examining Howard's ideals. In 1899, the Garden City Association was formed, and in the following year the Garden City Limited Company was established to raise capital for the development of a first Garden City.[3] The Garden City Pioneer Company, Limited was established on July 16, 1902 with an impressive group of directors including Howard, Ralph Neville, Edward Cadbury, and various other businessmen. The company secured its initial capital, through the sale of shares, within four months of issuing its prospectus; the Pioneer Company was wound up at the end of 1903. In the same year the Letchworth property was acquired in Hertfordshire with the purchase of a number of agricultural estates, and on Oct. 9, 1903 the first Garden City was declared under the First Garden City Limited company. This commenced the design,

---

[1] See, for example, K.C. Parsons and D. Schuyler, *From Garden City to Green City: The Legacy of Ebenezer Howard* (Baltimore: Johns Hopkins University Pres, 2002). p. 78; S.V. Ward, ed., *The Garden City: Past, Present and Future* (London: E & FN Spon, 1992); "The Garden City Idea," *The Architectural Review*, Number 976 (June 1978).
[2] Lewis Mumford, "The Garden City Idea and Modern Town Planning," in Ebenezer Howard, *Garden Cities of To-Morrow* (London: Faber and Faber Ltd., 1960), p. 29.
[3] Mervyn Miller, *Letchworth: The First Garden City* (Chichester: Phillimore & Co. Ltd., 1989), p. 20.

construction, and promotion of the town and surrounding agricultural belt. With the development of Letchworth, beginning in 1903, Howard's ideals were translated into practice. Simultaneously with the securing of the land the "Engineering Committee" organized a survey of the site.

At the same time consultations were undertaken regarding the selection of the designers for the overall layout of the town. The following firms were interviewed and submitted schematic plans by January 1904: the noted Arts and Crafts architects W.J. Lethaby and Halsey Ricardo; the relatively unknown architects Barry Parker (1867-1947) and Raymond Unwin (1863-1940); and a local architect based in Hitchin named Geoffrey Lucas.[4] Following a review of the three schemes, the plan submitted by Parker and Unwin was accepted by the Board on February 11, 1904.

The awarding of the Letchworth commission to Parker and Unwin would significantly enhance the reputation of the practice, which had been established in 1896, a partnership that would last until it was dissolved in 1914. Parker was born in Derbyshire, and, prior to joining with Unwin, was trained in various studios. Unwin was born in Yorkshire and grew up in Oxford. In 1884 he returned to northern England, where he became involved in various socialist movements. In particular, Unwin was strongly influenced by William Morris (1834-1896) and Edward Carpenter (1844-1929). Morris was the inspired and influential leader of the Arts and Crafts movement, while Carpenter was a well-known socialist poet and philosopher who Unwin knew personally, and who promoted the virtues of a rural living that combined manual, intellectual, and artistic pursuits. After the dissolution of the practice, Unwin would become an international champion of Garden City planning and design, responsible for town plans, publications, and architectural designs. The Parker and Unwin office began by designing mainly houses in the Arts and Crafts manner, as a result of this the partners were very involved in domestic design theory as was evident in their writings and practice. In 1902 they were asked to design the model village of Near Earswick, near York, for the Rowntree's, and in 1905 the firm was hired to plan Hampstead Garden Suburb in London, based largely on their experiences at Letchworth.

In their scheme for Letchworth, Parker and Unwin rejected Howard's geometrical diagram, and under the strong influence of Morris, developed a plan inspired by the medieval village, and what they believed to be its "organic unity."[5] Nevertheless, while responding to the specifics of the site selected for Letchworth, they maintained the basic tenets of Howard's vision. They also imposed a largely consistent architectural style derived from Arts and Crafts design and devoted much effort to the design and building of commodious housing. The original 1904 plan shows the general organization of site bisected by the railway. Areas adjacent to the station were laid out for shopping, the Town Square, and various public buildings. A sector was allocated for factory sites flanking the railway, with various sidings. Residential neighborhoods were carefully planned, interspersed with various kinds of green space from the large untamed Norton Common, to the narrow formality of Howard Park.

---

[4] Ibid., pp. 42-49.
[5] R. Fishman, *Urban Utopias in the Twentieth Century* (New York: Basic Books, Inc., 1977), pp. 67-68.

Notably the community design utilized the single-family house and garden throughout, maintaining relatively low densities, and a strong emphasis on green space.

Examining a drawing of the "The First Garden City—Plan of Estate and Proposed Town" published in June 1904 by The First Garden City, Limited, we can note the outline of the complete estate (3,818 acres), which would be expanded in certain areas in the future (for example, in 1912 land was purchased south of the Willian Road); the initial plan by Parker and Unwin showed a town area of about 1,200 acres. The estate was sandwiched between the small towns of Hitchin and Baldock, and incorporated the villages of Letchworth, Willian, and Norton, and part of the village of Radwell. The agricultural belt was implied, but not precisely defined. A plan published in 1913 shows the outline of estate and of the town,[6] with the near final outline of the complete estate (a parcel on the west edge would be later added). Over time there would be pressure to extend the size of the town by reducing the area of the agricultural belt, as indicated in a plan published in 1952. By 1949 the total estate area was 4,598 acres, with a total town area of 2,182 acres, for a projected population of 35,000.[7] Regardless, the original plan by Parker and Unwin has been closely adhered to since the establishment of the town.

As noted above, both Parker and Unwin wrote about architecture, mainly domestic design. Unwin also wrote extensively on the emerging discipline of town planning. In his architectural writings Unwin precisely addressed the design of the home with respect to planning, light, and air, and while he did not address the garden in great detail, it was inherent to his efforts to improve domestic environments. This is evident in his famous 1912 essay "Nothing Gained by Overcrowding!" in which he attempted to make an economic argument for lower densities, reduced roadways, and more gardens and recreation space.[8] Writing in 1909, in his text on town planning entitled *Town Planning in Practice* (the book was in part inspired by the writings of Camillo Sitte, who Unwin discovered after developing the Letchworth plan), in support of his concept of 12 units per acre, Unwin stated:

> Twelve houses to the net acre of building land, excluding all roads, has been proved to be about the right number to give gardens of sufficient size to be of commercial value to the tenants—large enough, that is, to be worth cultivating seriously for the sake of profits, and not too large to be worked by an ordinary labourer and his family. There will, of course, be men who can work more, and men who can only work much less, but in laying out of the

---

[6] C.B. Purdom, *The Garden City: A Study in the Development of a Modern Town* (London: J.M. Dent & Sons Ltd., 1913), p. 117.
[7] C.B. Purdom, *The Building of Satellite Towns: A Contribution to the Study of Town Development and Regional Planning* (London: J.M. Dent & Sons Ltd., 1949), p. 91.
[8] See R. Unwin, "Nothing Gained by Overcrowding!" in W.L. Creese, ed., *The Legacy of Raymond Unwin: A Human Pattern for Planning* (Cambridge, Mass.: The MIT Press, 1967).

land there are sure to occur great varieties in the size of the individual gardens which will allow for these differences.[9]

Unwin suggested that this density has been proven by the firm's experiments at Bournville, Earswick, Letchworth, Hampstead, and elsewhere. The emphasis on gardens and related labor can also be noted, as well as the "commercial" aspects of the gardens. In the same text, in his discussion of "plots and placing of buildings," Unwin stressed the importance of well-oriented houses, that also created well-composed streets; according to Unwin good site planning must take all this into account, including the basic planning of houses.[10] Further, Unwin was not enthusiastic about the use of fences between lots, particularly in the front gardens.

The history of the founding of Letchworth has been comprehensively documented by a variety of authors.[11] When built it was internationally recognized as a significant development in town planning, despite being connected with various historical precedents. The founders of the town, including Ebenezer Howard, emphasized the concept of a healthy town permanently surrounded by countryside, a community based on cooperation, and the importance of productive labor. The relatively inexpensive land that was acquired, combined with the ideals of the movement, inspired the low-density design of the town with its emphasis on green space and gardens. Ultimately, Parker and Unwin adapted the theories of Howard and others, by placing emphasis on the factors outlined above. However, the development of the town was slow. Early work involved a detailed survey of the site, sourcing water, and studying the drainage. A temporary railway station was established in 1903, and the first roads constructed.[12] Soon building plots were established and the first houses erected. The first decade was a period of enormous struggle for those invested in the project, one that saw many setbacks, but it also represented the clearest attempt to realize the Garden City model.

The scheme by Parker and Unwin was originally designed to accommodate 30,000 residents on 1,250 acres, resulting in an average of 24 persons per acre; 2500 acres of the estate was for the agricultural belt. The site was bisected by an existing railway line, and the design needed to address drainage, transportation, functional organization, etc.; the construction of infrastructure on the site began in the summer of 1904, the first houses were also begun at the same time by various private builders. The commitment to avoid land speculation was stressed in early documents, and in leasehold agreements.[13] In 1905, a Cheap Cottages Exhibition was held at Letchworth which attracted a lot of interest, but also did some harm in that the cottages on display were badly built and poorly located, and the overall site was very undeveloped, leaving visitors with a poor impression.[14] Many of the original residents of

---

[9] R. Unwin, *Town Planning in Practice: An Introduction to the Art of Designing Cities and Suburbs* (New York: Benjamin Blom, Inc., 1971), p. 320.
[10] Ibid., pp. 319-359.
[11] See, for example, Purdom, *The Garden City*, and Miller, *Letchworth*.
[12] Purdom, *The Building of Satellite Towns*, pp. 57-58.
[13] Purdom, *The Garden City*, pp. 47-48.
[14] Ibid., pp. 49-51.

Letchworth were enthusiastic middle-class pioneers committed to the social ideals of the Garden City, they were typical of the time, however, the community quickly gained the reputation for attracting eccentrics. By 1907, after a boom, the population grew to 4300, and a number of businesses had located in the town; newer residents tended to be associated with the businesses and included a larger working-class element attracted by work in the various factories that were established in the town. The demand for churches, clubs, and amusements rose as a result.[15] After a decade the character of Letchworth had been established, and it would argued by promoters of the Garden City that it was not a suburban development, due largely to the full range of industries and employment available.

C.B. PURDOM AND A CRITIQUE OF THE FIRST DECADE

By 1913 enough of the town was built to determine the relative strengths and weaknesses of its many concepts. During the First World War Letchworth prospered as it took in 3,000 Belgian refugees and became a center of munitions manufacturing. As the town was developed it did so in a scattered manner, and by 1925 was still missing essential elements, for example, there was very little development north of the railway line, and there remained gaps in the urban fabric.[16] C.B. Purdom (1883-1965), the noted Garden City advocate and chronicler, stated that between the two World Wars, Letchworth

> continued to exhibit no spectacular signs of development and no original features. In fact, it began to show the unassuming characteristics of ordinary English country towns that depend on nature and chance. There was little or no fresh enterprise, but the financial position was consolidated...and the garden city company took its place as a sound and well-established land company.[17]

Purdom praised the original plan and the commitment of the architects, however, he noted that the original plan did not adequately accommodate industrial sites.[18] Purdom also suggested that the most innovative period in the development of Letchworth occurred during the first decade of its history, and after this there was a decline in visionary effort on the part of the enterprise. Purdom attributed the slow growth of Letchworth to a lack of capital and government support. The undertaking of Letchworth was faught with numerous challenges that compromised some of Howard's original ideas, however, the pragmatism inherent to the Garden City movement resulted in the establishment of Letchworth.

Purdom was an active participant in the early Garden City movement, holding positions at both Letchworth and Welwyn Garden City, he began as a junior clerk

---

[15] Ibid., pp. 53-54.
[16] Ibid., p. 62.
[17] Ibid., p. 82.
[18] Ibid., p. 43.

working for the company, and then, after a severe illness, was the first resident of Letchworth. In his autobiography he described his initial experiences in the following terms: "From the start my new life was an exquisite delight. I was in a practically deserted countryside, in a village of a dozen scattered houses, without store or public house, though there were two larger houses inhabited by people who regarded the advent of the garden city with enmity."[19] He provided detailed accounts of the people, including the early town managers Thomas Adams (1871-1940) and W.H. Gaunt, and events that occurred as the town was established. Purdom would take over accounting duties for the company, and eventually he wrote several books, including key works on the history of Letchworth, and on the concept of the "satellite town." In 1918 he became the assistant secretary of the Garden Cities and Town Planning Association, and he would hold various positions with the enterprise that developed the Welwyn Garden City. Beyond the Garden City movement, he was active in drama, edited the journal *Everyman*, and was a biographer of the Indian teacher Meher Baba to whom he was devoted.

Purdom comprehensively documented the first decade of the development of Letchworth in his book *The Garden City: A Study in the Development of a Modern Town*, published in 1913. In the book he provided detailed information on Howard's theories, and on the history of early Letchworth. There were many innovations in the design of Letchworth, including the "agricultural belt," the emphasis on residential gardens, and foregrounding the act of gardening. As Purdom stated, Letchworth was conceived around the experience of being in a garden,[20] at both the intimate level of the residential garden and at the communal level of the town. Further, the use of the agricultural belt, or greenbelt, as a wide boundary separating the town from the country and providing a protected zone for agricultural and cultural amenities, helped create the sense of a town engulfed in greenery. Purdom began his text by describing the historic influence of cities, and then briefly discussed the effects on both town and country that resulted from the Industrial Revolution and large population migrations to industrial cities. Recognizing that there were a series of precedents for the Garden City he provided sketches of Robert Owen's schemes, the Village Associations movement, Buckingham's proposal for a model town, Richardson's project for Hygeia, and various model industrial villages (New Lanark, Port Sunlight, and Bournville).[21]

In the text Purdom described the origins of the Garden City, he noted that Howard's publications and his ideas, were generally well received, and that very shortly after the publication of *To-Morrow: A Peaceful Path to Real Reform* an enthusiastic group of supporters came together to plan the development of a Garden City. Despite the formalism of his diagrams, Purdom argued that Howard was correct in focusing much of his writing on the logistics of creating an actual model Garden City. Ralph Neville, as chairman of the council, and Thomas Adams, as secretary, were appointed in 1901; both would be key figures in realizing Letchworth, and both

---

[19] C.B. Purdom, *Life Over Again* (London: J.M. Dent & Sons Ltd., 1951), p. 42.
[20] Purdom, *The Garden City*, p. 113.
[21] Ibid., pp. 1-16.

would leave in 1906.[22] Conferences were held, to develop support for the initiative, at Bournville in 1901, and Port Sunlight in 1902, and these attracted much attention.[23] Purdom stated that three factors gave credibility to the movement: 1) that it was "a legitimate and proper undertaking for private enterprise," 2) that it had a simple economic structure, and 3) the existence of model towns at Bournville and Port Sunlight were evidence of practical success.[24] The prospectus of 1903 described the advantages of the site, and the ideals of the community, these included:

> Firstly, the provision of hygienic conditions of life for a considerable working population. Secondly, the stimulation of agriculture by bringing a market to the farmer's door. Thirdly, the relief of the tedium of agricultural life by accessibility to a large town. Fourthly, that the inhabitants will have the satisfaction of knowing the increment of value of the land created by themselves will be devoted to their own benefit.[25]

The emphasis on health, the stimulation of local agriculture, and structuring of land rents are noteworthy. Purdom suggested that the lack of detail in Howard's scheme, beyond its simple economic model, ended up being a strength of the concept, in that it attracted a wide range of people and ideologies.[26] One shortcoming in Howard's vision was the lack of attention paid to the development and growth of the community, instead he presented a fully developed vision of a town. Describing Howard's vision, Purdom wrote, "Its formality, order, and completeness, its fearful up-to-datedness, its cold, clean, and new atmosphere, were, however desirable in theory, more than a little repellant in prospect."[27] Further, the concept was "elastic"[28] and the resulting design for Letchworth, according to Purdom, avoided many of the limitations inherent to Howard's diagrams.

Addressing the architecture of the first Garden City, in 1904 the Garden City Company issued a document entitled "General Suggestions and Instructions Regarding Buildings Other than Factories on the Garden City Estate" which provided general guidelines for the design and construction of buildings; these would prove to be difficult to implement in a systematic way. Writing in 1913, Purdom questioned the architectural approach at Letchworth when he wrote:

---

[22] Adams would go on to a career in planning in both Canada and the USA. In the early 1920s he was elected chair of the Advisory Group of Planners overseeing the development of a regional plan for New York City and surroundings. Adams' work won the plan would lead to disputes with Raymond Unwin, Ebenezer Howard, and Lewis Mumford who charged that he was being cautious and conservative in his approach. The debate surrounding the regional plan is described in detail in P. Hall, *Cities in Civilization* (New York: Fromm International, 2001), pp. 782-795.
[23] Purdom, *The Garden City*, p. 24.
[24] Ibid., p. 25.
[25] Ibid., p. 34.
[26] Ibid., pp. 37-38.
[27] Ibid., p. 38.
[28] Ibid., p. 40.

> The problem at Garden City is, then, the discovery of a new kind of town architecture. For this discovery endless experiment is necessary, but if each builder does the best that is in him, if thought and care be given to each separate piece of work, and if the general effect be not totally ignored, it is possible that in time the Garden City will get what it wants.[29]

Despite Purdom's criticism of the conservative architectural approach used at Letchworth, the Garden City would always be a repository for more traditional kinds of architecture. Purdom was particularly critical of the use of a Georgian architecture of the public buildings, and the often cheap and poorly executed houses. He railed against the speculative housing that was common to Britain at the time, arguing instead for home ownership and houses built by the owners; this, he argued, would allow for houses that met the specifics of a site, including sun, and the requirements of the owner. Regardless, he did support the residential architecture of the Garden City and championed the good furniture of the time. A number of houses in Letchworth were designed by such prominent Arts and Crafts architects such as M.H. Baillie Scott, Halsey Ricardo, and C. Harrison Townsend, while many of the more ordinary houses were designed by Parker and Unwin (who moved their office to Letchworth), and various other local architects. The houses tended to be efficient and drew from British vernacular traditions. There were also experiments in co-operative housing, which eliminated the private kitchen in favor of a communal dining room. Despite efforts by Howard, and the architects, to develop this form of housing, this did not succeed in the face of financial and practical realities, and only one such development, a quadrangle arrangement called Homesgarth, was attempted.[30] As an aside, the New Towns that followed from 1946, and which drew many of its ideas from the Garden City, would embrace modernism to a large extent, particularly at Harlow New Town.

Purdom addressed the role of gardens, green space, the rural (or agricultural) belt, and buildings in his account, and he described the cultural and recreational amenities available during the early years of Letchworth. Letchworth was successful in attracting a number of industries during the first decade, including printing works and a variety of manufacturers, these were usually attracted by low rents, proximity to London, and the healthy living conditions for workers.[31] Further, Purdom tackled the question of housing for workers; at Letchworth this was typically cottages that strove to meet the needs of the working class in an economical manner. The new cottages in Letchworth often had to compete with less expensive housing in surrounding towns.[32] A major objective of the movement was to provide good living conditions for inhabitants, discussing infant mortality Purdom cited statistics that indicate that the rate at Letchworth was about half the national average, demonstrating that by 1913 the objective of a healthy community was achieved. On these issues he wrote:

---

[29] Ibid., p. 73.
[30] Fishman, *Urban Utopias in the Twentieth Century*, pp. 70-71.
[31] Purdom, *The Garden City*, pp. 140-153.
[32] Ibid., p. 161.

> Instead of fifty to eighty houses to the acre, in Garden City we have a maximum of twelve; instead of insanitary, inconvenient, and sunless dwellings, we have cottages with gardens; instead of crowded streets to play in, the children have the fields. In Garden City the home gets a chance of fulfilling its highest functions, of becoming a centre for refreshment, education and social life. All that is squalid, and mean is abolished.[33]

This underscores the idea that good education and general health were essential aspects of the Garden City, and that the inexpensive land obtained for the development of the town, allowed for better quality housing and amenities. In the concluding chapter to his book Purdom expressed optimism about what had been accomplished by 1913, and for the future development of Letchworth. In the Appendices various key figures in the early development of Letchworth provided essays, including Unwin and Howard. In his essay "How Far Have the Original Garden City Ideals Been Realized," Howard suggested only "to a small degree,"[34] and he expressed disappointment with the financing, the rent system, the shopping precinct, and the relatively few industries located in Letchworth at the time. In his contribution, H.D. Pearsall, the Chairman of the Howard Cottage Society, Ltd., addressed the state of worker's cottages, and noted that a minimum set of standards had been determined and met for each house.[35] Finally, one of the important results during the early development of Letchworth was the establishment of the Garden Cities and Town Planning Association, which would inaugurate the field of town planning in Britain.

CONCLUSION: GARDEN CITY INNOVATIONS

There is no doubt that the early Garden City was in many ways innovative; a number of proponents and historians of the movement have attempted to define those aspects of the model. For example, describing the Garden City pattern, the advocate F.J. Osborn summarized the Garden City in the following terms:

> towns of limited size, of controlled density, on a background of safeguarded countryside; towns where people live near work, with planned industrial zones; towns which are real communities of all classes, fully equipped for social life; towns with gardens and open space; towns in which architectural control aims at harmony in diversity. And as a means to these ends, unified landownership of large areas, with leasehold control to maintain good planning.[36]

---

[33] Ibid., p. 169.
[34] Ibid., p. 289.
[35] Ibid., pp. 262-265.
[36] F.J. Osborn, *Green-Belt Cities: The British Contribution* (London: Faber and Faber Ltd., 1946), p. 48.

This succinct description, by a key proponent of the Garden City, sums up its essential components. Complementing this description Osborn listed what he believed to be the main innovations of Howard's Garden City as: 1) planned dispersal, 2) limit of town-size, 3) amenities, 4) town and country relationship, 5) planning control, 6) neighborhoods, 7) unified landownership, and 8) municipal and co-operative enterprise.[37] All of these fall into general categories of planning and administration, and underscore Howards's key concepts. Further, Mervyn Miller, who has extensively documented the history of Letchworth, suggests that the key innovations of the enterprise were the simplicity and directness of Howard's ideas, the broad appeal of the concepts, the use of existing techniques, the employment of professionals in its early execution, and the persistently pragmatic approach of the founders.[38]

The historian Walter L. Creese, in his detailed study of the Garden City and its influence, also has identified a number of key ideas, including: using the village as a model for the town; implementing the picturesque design of townscapes; the density of the town; the careful design of road systems (for example, the use of cul-de-sacs); and the attention to site.[39] Creese notes that Unwin, in particular, used the village as a social and formal model for town design, was committed to understanding the past and championing the Middle Ages, and emphasized beauty in designing communities.[40] Creese also identifies specific innovations for the Garden City movement, beyond Parker and Unwin's concept of "twelve houses to the acre" and "nothing gained by overcrowding"[41]; these include "the side path and internal circulation, the cul-de-sac, the superblock, the curving street, the tree-lined verge, the closing and opening vista, the varied setbacks, climatological and focal siting, and any number of other refinements in the disposition of houses and buildings within a given setting."[42] He then outlines four general benefits of Garden City design:

> (a) It called attention to the ground as a distinct entity in itself, not merely a quantity waiting to be "improved" by structures; (b) it meant that the sequence of planning, from house to the community, to the region and thence the whole country could seem a more apposite process and that the planner's skill ought to be increasingly needed as the units enlarged; (c) it proved that the petty tyranny of the street over the home, or the opposite, as happened in early Leeds, could be restrained if other elements of urban composition could be better interpreted; (d) it indicated that visual planning was at its best when it was understood as an adjustment of solids, voids, edges, planes, and directions.[43]

---

[37] Ibid., pp. 32-33.
[38] Miller, *Letchworth*, pp. 210-211.
[39] Walter L. Creese, *The Search for Environment: The Garden City: Before and After* (Baltimore: Johns Hopkins University Press, 1992), pp. 169-173, 181.
[40] Ibid., pp. 169-173.
[41] See Unwin, "Nothing Gained by Overcrowding."
[42] Creese, *The Search for Environment*, p. 181.
[43] Ibid., p. 181.

Surprisingly none of these authors underscore one of the most important innovations of the early Garden City, the "agricultural belt" (later to become the greenbelt).

As Miller notes, the first decade of Letchworth was marked by a strong spirit of pioneering which is reflected in the design of the town and the type of people drawn to it, many of whom were free thinkers searching for an alternative life.[44] By the end of the first decade Letchworth supported a diverse population, the census of 1911 indicated a population of 5,324, with that number nearly doubling by 1921; even by 1961 the census indicates that the population was 25,515, below the designed number.[45] The town continued to struggle financially. As foreseen by Howard land rents were paid to the company, but shareholders were to receive their first small dividends in 1913, and the enterprise would never be financially successful.[46] By 1913 the structure of the town was established, and a core of businesses had located there. However, the outbreak of World War I would lead to decades of slow and difficult growth. In 1962 the Letchworth Garden City Corporation Act was passed by the British parliament, resulting in the transference of the town from a private enterprise, to one under public jurisdiction.[47]

The experimental nature of the initiative would be reflected in the desire for a self-sustaining community for both work and leisure. Green spaces, ranging from private gardens to the surrounding green belt, were to be used for the production of suitable food. Efforts to use the agricultural belt for farms and small holdings that would provide food to the town largely failed due the lack of capitalization; private gardens and allotments proved to be much more successful at growing food.[48] Nevertheless, the development of the agricultural belt, what has become widely known as the "greenbelt," would result in the development of new kinds of urban green space typologies, distinct from the isolated urban park. The Garden City placed particular emphasis on gardening and the gardener, and on revitalizing an integrated role for farming and the farmer. Therefore, the gardener and the farmer became two vital urban figures in the Garden City movement, figures not typically associated with urbanization.

---

[44] Miller, *Letchworth*, p. 88.
[45] C.B. Purdon, *The Letchworth Achievement* (London: J.M. Dent & Sons Ltd., 1963), p. 113.
[46] Purdom, *The Building of Satellite Towns*, pp. 147-172.
[47] Purdom, *The Letchworth Achievement*, pp. 56-104.
[48] Miller, *Letchworth*, pp. 136-141.

Ecologies of the Early Garden City

CHAPTER 7

Gardeners and Gardens

The garden is a place of pleasure, the *locus amoenus*, filled with joy, but it resounds in love laments of poets; is a place of feasts, entertainments of friend, a place according to Boccaccio, of sexual and intellectual freedom, a setting for philosophical discussions, and a restorative for both the body and the soul. It is a measured and well-ordered model of the universe, an experiment in immortality, a never-ending apparition of spring. It assumes the function of a picture gallery, a pinacotheca, a horticultural encyclopedia in vivo, a center of botanic and medical research, a theatre of imitation, competing with nature on nature's terms and conditions. Finally, it is a perpetual source of moral instruction.[1]

Traditionally, and fundamentally, the garden is a bounded piece of land devoted to the cultivation of plants, however, since its inception the garden has accumulated a diverse set of meanings and functions. The garden, and the gardener, emerged with the development of agricultural and urbanized cultures, as various pre-historic peoples abandoned, or were forced to abandon, hunter-gatherer modes of existence. For example, gardens were important in ancient Mesopotamian, Greek, Roman, and Aztec cities.[2] The garden blends horticultural practices derived from farming, with the cultural or artistic aspirations of an urbanized society, often as a place of escape from the travails of the city. The garden is a complex place, it is a place of work, pleasure, recreation, entertainment, philosophy, art, science, love, education, and symbolism; the garden is often described as a paradise. Michel Foucault has described "heterotopic" spaces as those that are other, that lie "outside all places."[3] According to Foucault, heterotopias have a precise role in society, and various defining characteristics. One of these is the power of "juxtaposing in a single real place different spaces and locations that are incompatible with each other."[4] Foucault invokes the garden as representative of this quality.

The confluence of culture (city/urban) and agriculture (country/rural), that occurs in the garden, or the space of cultivation, underscores the fact that the garden is a product of urbanization, so it is paradoxical to describe urban gardens and parks as the antithesis of the city. Octavio Paz reminds us that there are direct links between

---

[1] E. Battisti's definition of the Renaissance garden, quoted in Roy Strong, *Garden Party: Collected Writings 1979-99* (London: Frances Lincoln, 2000), p. 14.
[2] See K.R. Jones and J. Wills, *The Invention of the Park: From the Garden of Eden to Disney's Magic Kingdom* (Cambridge, UK: Polity Press, 2005).
[3] M. Foucault, "Of Other Spaces: Utopias and Heterotopias," in N. Leach, ed., *Rethinking Architecture: A Reader in Cultural History* (London: Routledge, 1997), p. 352.
[4] Ibid., p. 354.

cultivating the land and culture when he writes: "cultivating land means tilling it, working it to make it fruitful. Cultivating the mind or imparting culture to a people means improving them so they will bear fruit." [5] Paz's statement affirms the fundamental relationship between gardens and cities, and the abiding connection to agriculture and culture. Paz also suggests that it is the gardener who is the agent of this cultivation, the figure who plants, nourishes, and benefits from the garden, and that it is in the garden that the aspirations of an urbanized society are realized. As is evident from the long history of gardens, the garden unites town and country, the cultural and the earthly.

THE QUALITIES OF THE GARDENER AND GARDENING

The history of the garden is a testimony to its enduring qualities. Many authors have written about both the practical aspects of the garden (and gardening) and the cultural role that gardens have played, and the evolving design of gardens. An examination of some of the books written about gardening leads to a number of common themes.[6] Gardening is typically portrayed as a solitary and reflective activity,[7] and gardens are understood as places of craft, pleasure, contemplation, healing, celebration, and creativity. The gardener balances working in the garden, with an accumulating knowledge of horticultural practices, and a continuous struggle with nature representative of the human condition. Thomas C. Cooper notes that the "constitution of a gardener is a mercurial mixture of imagination and optimism."[8]

Using the model of the seasons, so often found in texts on gardens, the following general observations can be made. During the dormant winter months gardeners tend both to celebrate the accomplishments of the past seasons, ensure the garden is well protected from the harshness of winter, and dream of the promises of the future. While for many, particularly in the northern climates, the winter months are quiet, a time of deep contemplation and maintenance, in warmer climates the garden lives year-round. Many writers stress that gardeners are resilient, and that they are dreamers constantly re-imagining and planning for the spring months. Throughout the fitful spring months, the cultivation of the earth is truly underway with the reemergence of greenery, and the planting of new vegetation; the spring also provides opportunities for reflection particularly during the periods of unstable weather common to this season.[9] The mutability of the world is very evident during the spring months. Living with a degree of uncertainty, with the unfulfilled dreams of future possibilities, is

---

[5] Octavia Paz, "The Verbal Contract," in *Convergences: Essays on Art and Literature* (New York: Harcourt, Brace, Jovanovich, 1987), pp. 143-44.
[6] See, for example, Gertude Jekyll, *Gertrude Jekyll on Gardening* (Boston: David R. Godine Pub., 1984); Karel Capek, *The Gardener's Year* (New York: The Modern Library, 2002); H.J. Massingham, *This Plot of Earth: A Gardener's Chronicle* (London: Collins, 1944); Vita Sackville-West, *In Your Garden* (London: Michael Joseph, 1951); Thomas C. Cooper, *Odd Lots: Seasonal Notes of a City Gardener* (New York: Henry Holt and Company, 1995); Michael Pollan, *Second Nature: A Gardener's Education* (New York: Grove Press, 1995); Diane Dreher, *Inner Gardening* (New York: Quill, 2002).
[7] Dreher, *Inner Gardening*, p. 5.
[8] Cooper, *Odd Lots*, p. 9.
[9] Dreher, *Inner Gardening*, pp. 62-63.

something that all gardeners are familiar with. During the summer months, gardeners require both patience and presence.[10] The sensual aspects of the garden are very strong during the summer months. The many fruits of a garden are very evident during this season, and a particular "mindfulness" is necessary as one begins to harvest a garden.[11] Gardeners must be attuned to their plants in order to know when to properly harvest them. The changing seasons are underscored during the autumn months; this is the harvest season and the time for celebrating the harvest. It is a time of perseverance, a quality all gardeners possess.[12] The harvests from a garden initiate various kind of celebrations, from the personal to the shared. Throughout this season trees and shrubs are planted so that there are also new beginnings. It is also the time to clean up and protect the garden for the winter months.

In his book *Second Nature: A Gardener's Education*, the contemporary American writer Michael Pollan provides a meditation on the relationship between humans and nature that is the basis for all gardens. He begins by stating that a garden is rooted in a specific place, he notes that gardeners accept contingency, and that they tend to the practical over the theoretical. Pollan notes that despite the anthropocentrism of gardening, gardeners tend to also be broadminded and enlightened, accepting wildness and unpredictability, and the interconnectedness of living organisms.[13] Unlike the naturalist, gardeners are not romantic about nature.[14] Pollan suggests that gardeners maintain a 'quarrel' with nature, and that this has been the basis for human urban cultures.[15] He states:

> The gardener in nature is that most artificial of creatures, a civilized human being: in control of his appetites, solicitous of nature, self-conscious and responsible, mindful of the past and the future, and at ease with the fundamental ambiguity of his predicament—which is that though he lives in nature, he is not strictly of nature. Further, he knows that neither his success nor his failure in this place is ordained. Nature is apparently indifferent to his fate, and this leaves him free—indeed, obliges him—to make his own way as best as he can.[16]

Pollan suggests that the anthropocentric nature of the garden means that as an operational structure the garden as a model has wide application. He argues that gardening can improve a piece of land, often creating diverse and rich eco-systems, that human interventions can be ecologically beneficial by imitating nature's principles.[17] The gardener is a creator of ecologies. In this sense the garden could provide a powerful paradigm of how urban cultures can operate, and how we can

---

[10] Ibid., pp. 91-92.
[11] Ibid., pp. 111-13.
[12] Ibid., pp. 178-79.
[13] Pollan, *Second Nature*, p. 192.
[14] Ibid., pp. 192-93.
[15] Ibid., p. 193.
[16] Ibid., p. 196.
[17] Ibid., pp. 193-95.

move forward as humans in addressing the many ecological challenges of contemporary globalism.[18]

This notion is underscored by the work of the British gardener Jennifer Owen who has studied the ecological aspects of a large suburban garden in Leicester, England over a fifteen-year period (1972-1986) and determined that such a garden can produce a diverse eco-system. During the study she recorded 1782 species of animals, mainly insects, and 422 species of plants in the garden, she writes:

> A typical garden is a complex mosaic of tall and low vegetation, of open spaces and shade, forming an intricately structured and extremely patchy environment. Contrived plant diversity and permanent succession are thus combined with extreme structural heterogeneity. Within a small area there are elements of meadow, woodland and other habitats, so closely juxtaposed that a garden can be considered as a system of ecotones. There are edges everywhere, between herbaceous borders and lawn, shrubbery ad path, vegetable patch and compost heap, and so on.[19]

Owen underscores that gardens are complex artificial ecologies, and that even in a relatively small garden, it operates like larger ecologies influenced by the organization of the land, and the inherent boundaries contained within.

As the British ruralist H.J. Massingham, writing mainly in the 1930s and 1940s, pointed out, a garden is "the most fertile and productive place in the world,"[20] and it is in the garden where true meaning of cultivation has been preserved, as this has been generally lost in farming. Gardens bring together a diverse range of elements and people into a creative laboring. According to Massingham the true purpose of a garden is to produce a pleasurable kind of work, he wrote:

> Year in and year out it is nothing but work from morning to night. But it is work which continually sees the end in the means to achieve it. It satisfies the two primary needs of man, to nourish his body and delight his spirit. It is a perennial exercise of craftsmanship and freedom of choice and is the very base of reality.[21]

Ultimately, Massingham saw gardening as a fusion of work and play. Further, he wrote that the gardener acts as a supervisor of a garden's fertility, "the medium for this circulation between life and decay and renewal, nothing but an agent to secure its smooth functioning."[22] When a gardener takes his or her regular walk around their garden, it,

---

[18] Ibid., pp. 190-91.
[19] J. Owen, *The Ecology of a Garden: The First Fifteen Years* (Cambridge: Cambridge University Press, 1991), p. 349.
[20] Massingham, *This Plot of Earth*, p. 8.
[21] Ibid., p. 51.
[22] Ibid., p. 89.

is something between a survey, a diurnal stocktaking, a strategical campaign, an examination paper and a ritual. His emotions, too, are a blend of satisfaction, anxiety, speculation, disappointment and pleased surprise. He is at once a *paterfamilias*, a purging dictator, a doctor going his rounds and a silent psalmist.[23]

Describing the many kinds of disciplines inherently involved in gardening, Massingham wrote:

A garden is many things, a piece of man but also of nature, a text-book of economics, a stronghold of liberty, a means of safeguarding the principle of person, a vent of craftsmanship, a nurse of character, a condition of health and physical maintenance, a poem and a prayer: in brief a way of keeping body and soul together.[24]

The nineteenth century American gardener Charles Dudley Warner, in his book *My Summer in a Garden*, praised the soil and discussed the productive aspects and morality of gardening. He argued that the principal value of a garden is to teach the gardener "patience and philosophy, and the higher virtues the garden thus becomes a moral agent, a test of character. As it was in the beginning."[25] The garden teaches the great lessons of life, along with giving pleasure, and being a place of production. Warner was enamored with military metaphors, which are scattered throughout his book. In particular he described gardening as a battle with Nature, especially against weeds and pests.

In his account of gardening the twentieth century Czech writer Karel Capek, like Warner, suggested that gardeners primarily cultivate the soil.[26] Creating ideal soil involves digging, hoeing, and many other gardening tasks, along with creating the right mix of earth, compost, and the like. Capek wrote, using food metaphors:

A good soil, like good food, must not be either too fat, or heavy, or cold, or wet, or dry, or greasy, or hard, or gritty, or raw; it ought to be like bread, like gingerbread, like a cake, like leavened dough; it should crumble, but not break into lumps; under the spade it ought to crack, but not to squelch; it must not make slabs, or blocks, or honeycombs, or dumplings; but, when you turn it over with a full spade, it ought to breathe with pleasure and fall into a fine and puffy tilth. That is a tasty and edible soil, cultured and noble, deep and moist, permeable, breathing and soft."[27]

---

[23] Ibid., p. 136.
[24] Ibid., p. 279.
[25] Charles D. Warner, *My Summer in a Garden* (New York: AMS Press, 1971), p. 8.
[26] Capek, *The Gardener's Year*, p. 23.
[27] Ibid., pp. 99-100.

From this spring the garden. Capek also noted that along with the passion and eternal optimism of gardening, comes patience and a certain virtuousness.[28] Like the farmer, the gardener, is intimately involved in cultivating a piece of the earth. The maintenance of a garden also involves watering, weeding, pruning, and controlling pests and disease. As S. Reynolds Hole states, a garden brings something new every day to the gardener, there "is always something worthy of his care and admiration, some new development of beauty, some fresh design to execute, some lesson to learn, some genial work to do"; the labor of gardening produces a gracious and healthy happiness.[29]

Gardening is generally considered a morally pure activity, because through cultivation it provides instruction in productive labor, and instructs in the seasons and the cycles of life. It is an activity that attempts to manage a piece of land in a sustainable and ethical manner through disciplined hard work. Attuned to the weather and the seasons the gardener plants, nurtures, and reaps the benefits of his or her labor. All of the attributes of and actions associated with a garden anchor the gardener to a piece of land, to the rhythms of nature. The urban gardener contributes to a greater ecological awareness for those living in cities. The gardener, who occupies the middle ground between country and city, has been a vital figure throughout urban history, toiling to create and maintain islands and networks of greenery within the structure of cities. With the radical broadening of green space systems, since the mid-nineteenth century, we see that the gardener has had tremendous impact on the city. And yet, gardening is largely a solitary (although it can be performed in teams), or anonymous activity, and the gardener remains a mysterious, or shady, figure.

As an agent, the residential gardener typically operates in a small and enclosed space, or allotment, and tries to harmonize with the weather, the cycles of life, and the inevitable forces that conspire against him or her (storms, pests, drought, etc.). The gardener, as an agent of production, is a relatively solitary figure continuously reflecting on the possibilities a garden presents. Paradoxically, this engenders a larger awareness of the operations of ecologies and societies. The garden as a complex and productive cultural and horticultural space emerged at the same time as the city; it is situated between the farm and the square and resonates with creativity, freedom, innovation, and cultivation; urban gardens and parks blend individual, bureaucratic, and self-ordering types of agency.

THE GARDEN AS AN ASSEMBLAGE

As an assemblage, the garden has material, expressive, and territorial roles that combine in continuously changing ways. The material aspect of assemblage theory according to Deleuze and Guattari, is expressed in machinic terms, and involves bodies, actions, passions, and technologies while the expressive aspect addresses

---

[28] Ibid., pp. 81-82.
[29] S. Reynolds Hole, *A Book About the Garden and the Gardener* (London: Thomas Nelson & Sons, 1892), p. 34.

language, actions, and statements involving the collective.[30] Therefore, the material, or machinic, aspect of gardening organizes the physical and metaphysical labor that gardens require. All gardens are productive assemblages in that they are unpredictable, the gardener creatively operates within fields of continuously changing forces and processes, much like a painter. As an assemblage a garden unites a wide range of factors, players and spatialities into a changing set of intensities. As an assemblage of elements, forces, organisms, spaces, and agents, gardens have established a long history within certain constants. Following from assemblage theory, we can state that a garden is an evolving collection of organisms, practices, objects, effects, affects, expressions, responses, and the like. A garden is also a territoriality or set of territorialities.

Gardens are part of a region, they depend upon local factors, but are also created, or artificial, ecologies. They bring together a complex set of forces (soil, water, climate, nutrients, organisms, spatialities, structures, etc.) under the careful management of a gardener. A garden requires a great deal of attention and labor, and can produce many things, often engendering a larger awareness of the operations of ecologies and societies. While gardening is often described in militaristic terms, as a battle against agents of destruction, it more resembles a contemplative activity involving the creation of artificial ecologies, or assemblages, of plantings, spaces, and temporalities. Gardens, are typically precisely bounded spaces, however there is always the potential to interconnect gardens into patchwork, or smooth space, matrices.

As an assemblage a garden unites a wide range of factors, players and spatialities into a changing set of intensities. As ecologies gardens are effective, and they are also strongly affective, describing the sounds emitted by various plants in the wind, Gertrude Jekyll wrote:

> Nearly all trees in a gentle wind have a pleasant sound, but I confess to a distinct dislike to the noise of all the Poplars; feeling it to be painfully fussy, unrestful, and disturbing. On the other hand, how soothing and delightful is the murmur of Scotch Firs both near and far. And what pleasant muffled music is that of a wind-waved field of corn, and especially of ripe barley.[31]

The sonic qualities of a garden are extended by the songs of birds; the presence of animals in a garden also provides vitality. The territorial aspects of gardens must be expanded to include greater ecological opportunities, and the continual spatial transformations in gardens produced by seasonal and designed change is a paradigm for generating more complex urban models. The many elements that factor into the creation and maintenance of a garden, likewise, are essential to what it produces in terms of ecological effects, color affects, social and political opportunities, literary

---

[30] G. Deleuze and F. Guattari, *A Thousand Plateaus: Capitalism and Schizophrenia* (Minneapolis: University of Minnesota Press, 1987), p. 88.
[31] G. Jekyll, "Home and Garden," in P. Robinson, ed., *The Faber Book of Gardens* (London: Faber and Faber Ltd., 2007), p. 218.

and aesthetic pleasures, food, interactions between a wide range of species, microclimates, compost, etc. The gardener manages an infinitude of possibilities and draws great pleasure from his or her labor. Contemporary cities, facing enormous ecological, social, and technological challenges can be modeled after the garden.

The radical notion, put forward by the early Garden City movement, that the city could become both a garden and a community of gardeners, is still a model for thinking about the contemporary cities, despite the notable failures associated with the model. The original model was driven by social and structural ideas for integrating agriculture (the country) and the city and was not primarily focused on ecological factors. The suburban qualities typically attached to the Garden City, do often overlook its accomplishments: development of the greenbelt and numerous other green space typologies, the focus on gardening as collective and spatial activities, and addressing labour practices and urban health.

THE EVOLUTION OF THE PARK

Initially a garden was a private and defined space attached to a palace, monastery, or house. In ancient cities the garden was removed from the public realm, limited in its access and impact on the everyday life of the city. A rare exception occurred in ancient Athens where a sacred grove, the Academus, was the public place of philosophical discourse in the shade of trees.[32] In the medieval city the garden was typically private, hidden behind walls, and yet the city was intimately connected to the country. The medieval city was a compact and dense urban form precisely related to its surroundings, a metaphorical garden, by the fortification walls. In some medieval cities there were common green spaces, such as the green in front of the cathedral at Wells in England, and the sporting grounds on the edge of London. Many utopian visions of the city since the Renaissance, have been based on, or incorporated gardens.[33] By the sixteenth and seventeenth centuries large royal gardens and parks (a larger and simpler form of the garden) in European cities such as Paris and London were increasingly made public, folding the garden into the public space of the city. During the nineteenth century large urban parks, modeled after the English picturesque garden, became an essential aspect of cities throughout the world. In the urban park, which is in many ways a broader and simpler version of a garden, many aspects of the garden are lost, however, many new functions are also gained. There

---

[32] K. Jones and J. Wills, *The Invention of the Park: Recreational Landscapes from the Garden of Eden to Disney's Magic Kingdom* (Cambridge: Polty Press, 2005), p. 37.

[33] For example, Thomas More writes eloquently of a city, modeled after London, and defined by gardens, in his text *Utopia*: "They cultivate their gardens with great care, so that they have both vines, fruits, herbs, and flowers in them; and all is so well ordered, and so finely kept, that I never saw gardens anywhere that were so beautiful and fruitful as theirs. And this humour of ordering their gardens so well, is not only kept up with the pleasure they find in it, but also by emulation between the inhabitants of several streets, who vie with each other," in F.R. White, ed., *Famous Utopias of the Renaissance* (New York: Hendricks House Inc., 1955), p. 46. See also C. Paul Christianson, *The Riverside Gardens of Thomas More's London* (New Haven: Yale University Press, 2005); Christianson argues that More's Utopia was effectively a depiction of London.

has been a very distinctive progression in the use and conception of urban parks, or green space systems, since the mid-nineteenth century.[34]

Public parks rapidly became vital spaces in the city.[35] A variation on the Renaissance garden and/or royal forest, parks they shared many fundamental aspects of the garden. With the advent of large public urban parks, such as Central Park in New York City by Frederick Law Olmsted and Calvert Vaux (opened in 1858), green spaces have been an increasingly vital aspect of cities around the world. The American historian Galen Cranz, in her monumental book *The Politics of Park Design*, has outlined the evolving phases of park design in American cities since the mid-nineteenth century. She (and Bolland) has determined that there are five phases[36]: The Pleasure Ground (1850-1900), the Reform Park (1900-1930), the Recreation Facility (1930-1965), the Open Space System (1965-?), and the Sustainable Park (1990-Present). These phases reflect changes to the functions, role, and design of parks during this period.

In the first phase, which coincides with the second half of the nineteenth century, early American public parks provided a respite from the dense and hard environments of the city. They were romanticized fragments of country, or nature, in the city.[37] Public parks were places for "unstructured pleasure," that could include strolling, sporting activities, watching events, or attending meetings.[38] Activities in the park were different from those elsewhere in the city, and provided "exercise, instruction, and psychic restoration."[39] Time in the park was an escape from the city, and provided many pleasurable benefits. Beyond the picturesque landscape design with its emphasis on naturalistic landscapes, furniture and other amenities supported the wide range of activities. Institutions such as museums, botanical gardens, zoos, observatories, and music halls might be located in the park.[40] Promoted by health reformers, the early large urban parks would leave a lasting legacy in many American cities.

By the turn of the twentieth century social reformers began to alter the purpose and design of urban parks. The Reform Park was a place for leisure activities designed to educate children, immigrants, and the working classes.[41] Structured play was to result in better citizens, with parks becoming smaller and more formal in their design. The reformist aspirations of the urban park movement were abandoned in the 1930s as a new pragmatism emerged that placed emphasis on providing sporting facilities that were often located in new suburban communities.[42]

---

[34] See G. Cranz, *The Politics of Park Design: A History of Urban Parks in America* (Cambridge, Mass.: The MIT Press, 1982); and G. Cranz and M. Bolland, "Defining the Sustainable Park: A Fifth Model for Urban Parks," *Landscape Journal*, Volume 23:2 (2004).
[35] See Richard Sennett, *The Fall of Public Man* (New York: Alfred A. Knopf, 1977).
[36] See Cranz, *The Politics of Park Design*; and Cranz and Bolland, "Defining the Sustainable Park: A Fifth Model for Urban Parks."
[37] Cranz, *The Politics of Park Design*, p. 5.
[38] Ibid., p. 7.
[39] Ibid., p. 8.
[40] Ibid., p. 14.
[41] Ibid., p. 61.
[42] Ibid., pp. 101-103.

By the 1960s "open space," influenced by the modernist city, became the new paradigm. The spaces were not programmed, intended to be fluid and open to a wide range of activities.[43] Often spaces ended up being featureless and functionless, part of a general trend towards the simplification of park design. As Cranz writes, "If the pleasure ground was an antidote, the reform park was a mechanism of social progress, the recreation facility a public service, and the open-space park a stimulant...."[44] Beginning in the 1990s public green space was harnessed to notions of creating more sustainable cities. This coincides with efforts to "green" the city, and to develop sustainable infrastructure.

There are many benefits provided by a park, as in a garden. Parks play an aesthetic role, an economic role, a sustainable role, a recreational role, a social role, and they affect tourism, livability, and health (physical and mental). Since the nineteenth century urban green spaces have tended to become maintained landscapes, these systems that have become larger, simpler ecologies, and managed by urban bureaucracies. Parks departments and hierarchies of employees, conceive of green space systems as large management problems, where regimes of planting, watering, mowing, fertilizing, weeding, etc. are systematically employed to support the functional and aesthetic programs these spaces fulfill. These have contributed little to the overall ecological effectiveness of cities, but this could be changed if tendencies towards the segregation of functions, spaces, and systems could be overcome.

CONCLUSION: BEYOND MAINTENANCE

Since the nineteenth century cities globally have increasingly incorporated a range of green space typologies from the private garden, comprehensive park systems, to encircling greenbelts. Each typology captures some aspect of the primal garden, these spaces have become part of the complex ecologies of cities and will increasingly need to be integrated as cities strive to find new ways of addressing the current environmental crisis. Like in the case of the Garden City movement, developed at the end of the nineteenth century, we can consider cities as either gardens, or complex artificial ecologies, or as a tapestry of patches, only some of which are gardens and parks. Nevertheless, cities are ecologies of infrastructure, buildings, people and other organisms, social and political systems, biological and environmental factors, etc. All of these are found in the garden, cultivated through the actions of a gardener, shaping forces and allowing for the element of chance. The gardener manages nature, does not control nature. Gardeners celebrate the vagaries of climate and weather, deriving a wide range of productive pleasures from their labors. Gardening demands responsibility. It is a continually evolving process as adjustments are continually made. But gardeners tend to be practical, they enjoy the constant cycle of labor associated with gardening but change in the structure of a garden tends to be slow and incremental. And yet, an experienced gardener is able to create a relatively complex micro-ecology.

---

[43] Ibid., pp. 135-138.
[44] Ibid., p. 232.

The garden as a complex and productive cultural and horticultural space is situated mid-way between the farm and the urban square, and while typically found in cities may also be found in rural landscapes (associated with villages, farms and villas). Gardens resonate with creativity, freedom, innovation, and the cycles of life from birth to death. A garden tends to increase fertility, and ecological effectiveness, relative to surrounding urban, farm, or wilderness lands. While, resolutely local and regional, gardens have always engaged with imported species and techniques, part of their urban legacy.

In urban environments where there is a high preponderance of gardens (suburbia, towns, some historic environments, etc.), then these helps ameliorate general environmental conditions: heat dissipation, absorption of pollutants, support of other species, habitat for numerous organisms, etc. The garden as a model for the contemporary city, as proposed by the original Garden City, suggests that cities could be conceived as complex and evolving ecologies, richer than the land prior to urbanization. Gardens are productive and are able to manage waste, they are constantly modified to respond to change. Gardens respond to seasonal and local factors and can be spatially open. Parks, which are not as ecologically productive as small residential gardens, nevertheless, also help with overall urban ecologies. The heterotopic nature of the garden, a space that can unite the country with the city needs to be reinvested. The modernist open-space or green space system, as Cranz notes, is now largely devoid of purpose.[45] The urban park needs to be redefined as a garden for the collective, and the role of gardener also needs reconsideration. The public park is enjoyed by the society at large as an aesthetic experience, a moral territory, a recreational zone, a public space, or part of the biology of the city, all of which are part of what comprises a garden.

Gardening, and agriculture have over the last century become increasingly industrialized, heavily reliant on machinery and a wide variety of fertilizers, pesticides, and herbicides. The private gardener, so essential to the structure and ethos of the Garden City, has been effectively replaced by the landscaper, someone more concerned with maintaining a visually pleasing landscape. The typical suburbanite uses a plethora of techniques and services to create perfectly maintained landscapes. In particular the ubiquitous lawn, with its own distinct cultural history, now often signifies that there is merely a garden-like space.[46] Suburban gardens have lost the many of the cultural qualities historically associated with gardens. While the pervasive sense of greenery, and the basic structural model found in the Garden City, has been retained, the ethos of the Garden City has been lost. The garden, actually and metaphorically, has been converted into a green space system that is more about appearance than any moral, civil, or existential potential, the ecological potential of the garden is largely negated. However, in recent years there has been a growing sense of the need for new ecologically responsible practices in cities. Ironically, the

---

[45] Ibid., pp. 135-54.
[46] See, for example, A. Wilson, *The Culture of Nature: North American Landscape from Disney to the Exxon Valdez* (Toronto: Between the Lines, 1991) and, G. Teyssot, ed., *The American Lawn* (New York: Princeton Architectural Press, 1999).

pervasive use of the garden as an urban model during the twentieth century has resulted in the loss of complexity historically associated with the garden.

There are various types of gardeners involved in cities, from the landscape architect, to the parks department employee, and the private gardener. The private gardener, as an agent, typically operates in a small and enclosed garden, or allotment, and is harmonized with the weather, the cycles of life, and accepts the inevitable forces that conspire against him or her (storms, pests, drought, etc.). Typically, the modern gardener has been part of a maintenance crew, or the individual homeowner laboring to keep up green spaces for both the benefit of themselves and the collective. In effect the gardener has become the missing figure in the development of the city in the twentieth century, which has derived so many ideas from the Garden City movement. The garden, a space that can unite the country with the city needs to be reinvested. The city as a garden, comprised of gardens, remains a powerful paradigm for the ecological, or sustainable, city. The gardener must reappear as a real urban figure, not an anonymous member of the maintenance crew.

CHAPTER 8

Farmers and Greenbelts

Cities comprise a broad range of agents and agencies, cities have always created classes and specialists: administrators, priests, business owners, police and military, educators, rulers, artists, etc. Cities have always integrated those who live in the city, with surrounding agricultural lands and extensive trading networks. The Garden City, with its inherent emphasis on gardens and uniting the city and the country, places import on two often overlooked agents, or urban figures: gardeners and farmers.

FARMS AND FARMING

William Conlogue defines agriculture as an activity that "bounds, arranges, and systematically transforms nature into something we can eat, wear, or otherwise utilize."[1] Conlogue recognizes that farming establishes boundaries on the land, and implies that it creates arrangements (assemblages) of practices, materials, agents, and land in order to produce. And yet, while farmers continuously confront uncertainty many of their practices are designed to overcome it. Describing agriculture as a process, J.A. Montmarquet has written:

> Agriculture enhances certain basic processes or organic development through the deliberate manipulation of environmental variables. Through such manipulation, forms of plant and animal life are transformed into useable products for human life. The extent of this manipulation is more extensive than mere hunting or gathering but cannot extend to the point that the farmer has become a mere technician or applied scientist.[2]

Here, Montmarquet acknowledges the impact that agriculture has on the land, and that it establishes ecologies for the growing of food, and that this activity cannot be reduced to merely a technical exercise. However, since the nineteenth century farming has become increasingly industrialized in both the techniques used, and the way it is organized.

Since the origins of agriculture, farming has been seen as promoting a set of basic virtues: "justice, honesty, independence, courage and a capacity for hard work."[3] There are various universal qualities associated with farming, which can be found in

---

[1] W. Conlogue, *Working the Garden: American Writers and the Industrialization of Agriculture* (Chapel Hill: University of North Carolina Press, 2001), p. 5.
[2] J.A. Montmarquet, *The Idea of Agrarianism: Form Hunter-Gatherer to Agrarian Radical in Western Culture* (Moscow: University of Idaho Press, 1989), p. 20.
[3] Ibid., p. 26.

the writings of the noted American agrarianist Wendell Berry, and which will be drawn from in order to examine some of the operational aspects of agriculture. Berry defines a farmer in the following terms:

> A competent farmer has learned the disciplines necessary to go ahead on his own, as required by economic obligation, loyalty to his place, pride in his work. His workdays require the use of long experience and practiced judgment, for the failures of which he knows that he will suffer. His days do not begin and end by rule, but in response to necessity, interest, and obligation. They are not measured by the clock, but by the task and his endurance; they last as long as necessary or as long as he can work. He has mastered intricate formal patterns in ordering his work within the overlapping cycles—human and natural, controllable and uncontrollable—of the life a farm.[4]

The rootedness to place is consistent with the history of urbanization, and the consequent striation of space. The dependence on experience and being able to respond to predictable and unpredictable forces are also inherent to cultivating land. The following statement by Berry is consistent with a certain kind of ecological agency: "the responsible consumer must also be in some way a producer. Out of his own resources and skills, he must be equal to some of his own needs."[5] This resonates with the emphasis championed by the Garden City movement on purchasing local agricultural produce and on making gardens places for growing food.

As Berry states "food is a cultural product,"[6] and that a healthy community "reveals the human necessities and human limits; the production of food clarifies our inescapable bonds to the earth and to each other."[7] He points out, by merely being consumers of food, as opposed to producers and consumers, we exist "as merely a conduit which channels the nutrients of the earth from the supermarket to the sewer."[8] This is an apt description of infrastructure systems, and how these bypass important ecological cycles. The production of waste, inherent to contemporary agriculture, has nothing to do with the original productive processes of agriculture. The sanitation, or "technological purification," of the body has resulted in the general pollution of environments.[9] Discussing the contemporary divide between urban and rural, Berry writes:

> cities exist in competition with the country; they live upon a one-way movement of energies out of the countryside—food and fuel, manufacturing materials, human labor, intelligence, and talent. Very little of the energy is

---

[4] W. Berry, *The Unsettling of America: Culture and Agriculture* (San Francisco: Sierra Club Books, 1986), p. 44.
[5] Ibid., p. 24.
[6] Ibid., p. 43.
[7] Ibid., p. 43.
[8] Ibid., p. 136.
[9] Ibid., p. 136.

ever returned. Along with its glittering "consumer goods," the modern city produces an equally characteristic outpouring of garbage and pollution."[10]

Berry's late twentieth century writings on the state of agriculture resonate with issues surrounding cities and agriculture at the end of the nineteenth century, issues that the Garden City movement attempted to address. In his essay entitled "A Defense of the Family Farm," which defends the role of the small farm, organized around the house and farmed by the inhabitants, Berry argues for the "responsible maintenance of the health and usability of the place."[11] Against the tendencies of industrial and corporate agriculture, the family farm depended on a long-term connection between the family and the land, it also demanded a commitment to laboring with the land, to produce a yield for both personal use and market. Farm living has provided, according to Berry, a healthiness and satisfaction[12] for those family farmers, and ultimately a sensibility informed by sustainability; importantly the traditional family farm created a supportive, if dispersed, community.

The writer and independent farmer David Mas Masumoto has written about the annual cycles of farming in an evocative manner. In his section on Spring he writes about the land and the importance of walking it:

> the farmer walking his fields can feel the changing landscape beneath his boots, he can sense the temperature changes with the different densities of growth and smell the pollen of blooming clover or vetch or wildflowers. He appreciates the precarious character of nature. As if running your fingers over a finely crafted quilt, you can feel pattern upon pattern.[13]

Like gardening, farming involves the manipulation, or "coaxing,"[14] of nature, often on a large scale; it is an aspect of the striation of space that urban cultures construct. Masumoto suggests that traditional family farming, as opposed to corporate agriculture, is a contemplative activity attuned to the seasons. Historically a farm has been a working home. Writing about the seeming stability of farming, Masumoto states: "My world may seem unchanged to casual observers, but they are wrong. I now know this: if there's a constant on these farms, it's the constant of change."[15] Change is rapid for farmers and gardeners, particularly that brought on by changing weather, however, there are numerous other kinds of change that are often too subtle to be observed by those who do not farm or garden. The ability to react to change and uncertainty, and the construction of artificial ecologies, inherent to both farming and gardening, provides opportunities for moving towards more sustainable environments.

---

[10] Ibid., p. 137.
[11] W. Berry, "A Defense of the Family Farm," in *Home Economics* (San Francisco: North Point Press, 1987), p. 162.
[12] Ibid., pp. 164-5.
[13] D. M. Masumoto, *Epitaph for a Peach: Four Seasons on My Family Farm* (New York: Harper, 1995), p. 11.
[14] Ibid., p. 47.
[15] Ibid., p. 183.

The nature of agricultural labor, as described by Berry, and Masumoto is virtually identical to the way that gardening has been typically described. These authors describe more traditional practices, that involved a direct and productive working relationship with the land, as opposed to the exploitive techniques often used in large-scale agricultural operations. Unlike gardening, farming is a business that involves intense work and employment, with the production of food as its primary result. They both involve the continuous working of land; as in any artificial ecology it is about establishing a dynamic balance between all the forces at play. Unlike a typical garden, a farm tends to produce a simple ecology, based on fields planted with a single crop, often with a loss of indigenous species.

Many writers on agriculture stress the importance of productive labor. They also note that these activities involve accumulating knowledge, and a creative element; they are creative forms of production. Peter Hallward, discussing the creative impetus in the work of Deleuze, writes:

> Deleuze presumes that being is creativity. Creativity is what there is, and it creates all that there can be. Individual facets of being are differentiated as so many distinct acts of creation. Every biological or social configuration is a creation, and so is every sensation, statement or concept. All these things are creations in their own right, immediately, and not merely on account of their interactions with other things.[16]

Farming and gardening can be understood as creative forms of labor, participating in the continuous making of assemblages of planting, seasons, practices, and spaces. This affirms the role of farmers and gardeners as creative agents participating in the large ecological forces that shape environments. Implied in this, is a critique of industrialized forms of agriculture. The Garden City, with its emphasis on local farmers providing for local markets, is a testimony of this; the civic and creative aspect of productive labor was at the heart of the movement.

With the advent of industrialized farming in the late nineteenth century there has been a steady erosion of the family farm, replaced by increasingly larger corporate farms. David Orr argues that a "new agrarianism" will need to emerge if environmental challenges are going to be addressed. In a manner reminiscent of the Garden City movement he suggests that there will be a return to local marketing of produce, farms will become smaller and more diverse, and the "new agrarianism will be powered by sunlight, not fossil fuels, and agriculture will become part of a larger strategy aimed at conserving soil and storing carbon the new agrarianism in some way will have to acknowledge that we remain hunter-gatherers by temperament."[17] Orr suggests that farming needs to be promoted as a worthwhile and enjoyable way of living. The continuous expansion of cities has resulted in encroachment on agricultural lands, and a blurring of the divide between city and country. However, as

---

[16] P. Hallward, *Out of this World: Deleuze and the Philosophy of Creation* (London: Verso, 2006), p. 1.
[17] D. Orr, "The Urban-Agrarian Mind," in E.T. Freyfogle, *The New Agrarianism: Land, Culture, and the Community of Life* (Washington, DC: Island Press, 2001), p. 97.

more agriculture is established within city limits and peri-urban conditions grow, the result will be a landscape with a tapestry of uses including residential, commercial, agricultural, industrial, and recreational uses, or a continuous patchwork of development that could result in effective ecologies, a condition reminiscent of Frank Lloyd Wright's Broadacre City project. The Garden City attempted to address many of these issues with bounded urban development, the common ownership of land, and rejuvenated local agriculture.

The plan for Letchworth Garden City does not go as far as integrating farming into the structure of the town, although the growing of food was intended to occur in private gardens and in the agricultural belt. However, the presence of farming was evident in the compact size of the town, and the proximity of the surrounding farms to all parts of the town. This was particularly the case in the first decade as the town developed in a relatively haphazard way. By 1913 the agricultural belt, which originally comprised two thirds of the entire land area, supported 75 tenants working parcels of land of varying sizes; the rates of return and condition of the agriculture were deemed "satisfactory."[18] As Purdom stated the Garden City "does not, like other towns, destroy rural pursuits; it intensifies them."[19]

THE FUNCTIONS OF GREENBELTS

In the nineteenth century the "agricultural belt" developed as a concept for dealing with the edge, to act as a mediating condition between town and country and limiting growth as a means for controlling land speculation. The agricultural belt was integral to Ebenezer Howard's vision for the early Garden City, which in harmony with a range of public and private green spaces was to perform a variety of functions. For Howard, the agricultural estate, or belt, supported agriculture, some institutions, natural preserves, managed waste, and provided a vital source of rent for the whole Garden City enterprise.[20] Describing this aspect of his vision, Howard wrote:

> All the sewage and other refuse of the town is utilized on the agricultural portions of the estate which is held by various individuals in large farms, small holdings, allotments, cow pastures, etc.; the natural competition for these various methods of agriculture, tested by the willingness of occupiers to offer the highest rent to the municipality, tending to bring the best systems of husbandry.[21]

The agricultural belt was also intended to give a precise edge to the town, to provide fresh air, act as a buffer from surrounding communities, and be used for the

---

[18] See H. Burr, "Agriculture and Small-Holdings in Garden City," in C.B. Purdom, *The Garden City: A Study in the Development of a Modern Town* (London: J.M. Dent & Sons Ltd., 1913), pp. 272-283.
[19] Purdom, *The Garden City*, p. 116
[20] Ebenezer Howard, *To-Morrow: A Peaceful Path to Real Reform* (London: Swan Sonnenschein & Co., Ltd., 1898), p. 24.
[21] Ibid., p. 17.

production of food.[22] Economically, it was a way of limiting the growth of the town, of controlling land speculation on the edges of the town, and of stabilizing the value of land in the surrounding agricultural area. C.B. Purdom wrote in 1949 that the agricultural belt "is a wide stretch of food-producing land surrounding the town retained as an integral part of the town's economy."[23]

The nineteenth century precedents for the agricultural belt, or "greenbelt," include the 1837 plan for Adelaide attributed to William Light, and James Silk Buckingham's scheme for a model town.[24] While not invented by Howard the agricultural belt was widely promoted by the early Garden City movement. Since the nineteenth century the concept has gained relatively wide application. Addressing Howard's original text and the concept of the agricultural belt Thomas Adams, an early administrator at Letchworth and then an advocate for town-planning in North America, suggested that the belt would not effect a new relationship between the town and country in and of itself.[25] According to Adams the agricultural belt, rather than the Garden City in general, was designed to support small agricultural operations and labor, he wrote that it was:

> 1st. —To convert as much land as practicable into small holdings.
>
> 2nd. —To encourage agricultural labourers to acquire small holdings, and the factory workers to cultivate allotments or gardens.
>
> 3rd. —To promote co-operation among the tenants.
>
> 4th. —To promote technical education and provide advice.
>
> 5th. —To give long leases and equitable conditions of tenure, and to encourage tenants to invest capital in improvements and housing accommodation.
>
> 6th. —To establish credit banks in order to give small holders financial assistance.
>
> 7th. —To encourage the development of suitable small industries in the villages.

---

[22] C.B. Purdom, *The Building of Satellite Towns: A Contribution to the Study of Town Development and Regional Planning* (London: J.M. Dent & Sons Ltd., 1949), p. 442.
[23] Ibid., p. 439.
[24] See F.J. Osborn, *Green-Belt Cities: The British Contribution* (London: Faber and Faber Ltd., 1946), pp. 167-80.
[25] Thomas Adams, *Garden City and Agriculture: How to Solve the Problem of Rural Depopulation* (Hitchin: Garden City Press, Ltd., 1905), pp. 38-39.

8th. —To provide up-to-date facilities such as water supply under pressure, siding accommodation, good roads, etc., for increasing the productiveness of the soil, and for promoting the rural industries.[26]

The objectives for the agricultural belt were never fully realized in practice, however, they describe the union of land, labor, infrastructure, and community that was the basis of the Garden City. Adams argued that gardens and allotments can "produce wealth," and that this is better than "dissipating both time and money in the public-house."[27] Consistent with the reformist agenda of the period, Adams wrote: "The marriage of town and country effected by Garden City will not only improve the physique of the factory worker, it will improve the intelligence and character of the rural labourer."[28] He cited examples of small holding and co-operative associations throughout England.

Adams suggested, following Howard, that the agricultural estate (or belt) be divided between large farms, small holdings or allotments, rural residences (with over two acres of land), and public institutions.[29] The most suitable types of agriculture would have been small dairy farms or market gardens. He discussed small holdings as both primary and secondary sources of income, and also described the role of allotments.[30] Adams wrote:

It will generally be found that a factory worker or artisan who is engaged in a profitable industry cannot afford and does not require from a health point of view, to cultivate more than an eighth of an acre. When ordinary farm land is first converted into garden land it requires patient, careful and skillful cultivation for two or three years.[31]

However, as Michael Simpson points out in his biographical study of Thomas Adams, the rural policy of the Garden City struggled from the beginning in light of general living and working improvements in Britain, and the failure of the small holdings movement.[32]

Many of the original inhabitants of Letchworth came from cities, and had little or no knowledge of rural life, however, it was hoped that they would develop a country sensibility. An example of this was W.G. Furmston, the manager of the Skittles Inn, who left a factory job in London, and with his family, moved to Letchworth as one of the first tenants. He learned to manage a productive small holding while maintaining his job and would be considered an ideal example of the Letchworth social and

---

[26] Ibid., pp. 43-44.
[27] Ibid., p. 48.
[28] Ibid., p. 51.
[29] Ibid., p. 87.
[30] Ibid., pp. 90-97.
[31] Ibid., p. 95.
[32] Michael Simpson, *Thomas Adams and the Modern Planning Movement: Britain, Canada and United States, 1900-1940* (London: Mansell, 1985), p. 26.

economic experiment.[33] While the agricultural belt supported a range of tenants and was strong in the production of dairy products, fruits, and vegetables, it did not truly integrate agriculture into the life of the town, nor provide the expected economic benefits that were hoped for in the original vision.

BENEFITS AND PROBLEMS

Ultimately, the use of the agricultural belt attempted to reorganize the traditional relationship between farmer and urbanite, and between city and country. The separation between city and country is addressed in a unified scheme, through the creation of a thick boundary condition designed to negotiate between the town and the surrounding agricultural land in a productive way. The agricultural belt also redefined the historic urban boundary.

Raymond Unwin, who would become an important proponent of town planning, was concerned with the visual beauty, or picturesque qualities, of townscapes, and commented on the ragged edges of modern towns, which he described as "that irregular ring of half-developed suburb and half-spoiled country which forms a hideous and depressing girdle around modern growing towns."[34] He argued for agricultural belts to define the edges of the town, and coherent gates on the major approach roads and railway stations. Further, he questioned the role of allotments and small holdings in the agricultural belt as he found these too often be unsightly, equating them with "shanties."[35] The multi-functional aspect of the agricultural belt suggests that it played an important role in the development of the Garden City concept. However, the agricultural belt or greenbelt has also become an ambiguous kind of space encircling many cities around the world. As the concept for the greenbelt developed it produced various problems, including increasing land costs, and dispersing development beyond the belt.[36] In the original vision of Howard's the agricultural belt played an active role in the life of the town, providing a vital interim layer between town and country. However, many subsequent greenbelts, while they conserve and preserve land, do not necessarily play this role.

The original agricultural belt was conceived as a fixed spatial element that acted as a permeable boundary, a habitat both cultivated and wild, a source of food, and a zone for managing waste. The bounding of the community was to provide a town and country interface, but most importantly, to put an end to land speculation, and the endless creep of cities, an anti-sprawl device. Further, as Robert Fishman states:

---

[33] See W.G. Furmston, "A Small Holding in Letchworth," in *Garden City and Town Planning Magazine*, Vol. 10, No. 8, (August 1920), pp. 176-179.
[34] Raymond Unwin, *Town Planning in Practice: An Introduction to the Art of Designing Cities and Suburbs* (New York: Benjamin Blom, Inc., 1971), p. 154.
[35] Ibid., p. 164.
[36] See Robert Freestone, "Greenbelts in City and Regional Planning," in K.C Parsons and D. Schuyler, eds., *From Garden City to Green City: The Legacy of Ebenezer Howard* (Baltimore: Johns Hopkins University Press, 2002), pp. 82-83.

The garden city was also limited in size in order to concentrate and intensify the life that took place within its limits. The garden city was not only an escape from the overcrowded, inhuman metropolis but also a new and higher locus of urbanity, a place where a genuinely urban complexity of activities could be carried out within a human-scaled container.[37]

Fishman argues that the bounding of the city opposed the early twentieth century tendency towards progress and expansion. Fishman also states that the model of a bounded community can avoid the high costs associated with the large metropolis, providing high quality environments and stable conditions for employers and workers alike.[38] This model has worked, but has been undermined by the common disconnection between employment and residence (often linked to the automobile), and the uncertainties of employment. Fishman suggests that the boundedness of the Garden City is a vital, and over-looked, aspect of its legacy. The agricultural belt also protected the amenities of the town, supplied food, and restricted expansion on the edges.[39] The agricultural belt, or greenbelt, concept claimed to check urban sprawl, dampen land speculation, stimulate urban infill, protect agricultural land and scenic resources, promote recreation, preserve town character, promote proximity to rural life, provide greenfield sites for particular institutions, and preserve wildlife and vegetation.[40]

However, Robert Freestone has recognized that the agricultural belt, or greenbelt, has generally restricted sprawl and protected agricultural land adjacent to cities, after analyzing various planning studies he suggests that there are the following problems associated with this approach:

1. Greenbelts increase land and house prices.

2. Greenbelts can protect land of average environmental quality.

3. Greenbelts increase car travel.

4. Greenbelts divert development deeper into the countryside.

5. Greenbelts increase development pressures within existing centers.

6. Greenbelts can have a range of unpredictable effects.

7. Greenbelts do not necessarily increase public access to nonurban land.

---

[37] Robert Fishman, "The Bounded City," in Parsons and Schulyer, eds., *From Garden City to Green City*, pp. 58-59.
[38] Ibid., pp. 60-61.
[39] Purdom, *The Building of Satellite Towns*, p. 447.
[40] Freestone, "Greenbelts in City and Regional Planning," p. 78.

8. Greenbelts are not always environmentally just.

9. Greenbelts are a negative and inflexible means of development control.

10. Greenbelts do not constitute a regional settlement strategy.[41]

These concerns address issues such as the "leap-frogging" of development beyond the greenbelt, lack of flexibility, and the tendency to increase land costs.

From the concept of the agricultural belt developed a whole host of new green typologies have emerged during the last century that inform contemporary urban design: greenbelts, green wedges, green webs, green corridors, and greenways. With respect to the agricultural belt, Purdom wrote, "in the garden city the agricultural belt is recognized, to some extent it is defined, and, so far as it forms part of the garden city, it belongs to the unity of the town and is preserved, maintained, and developed equally with all other parts of the town."[42] The agricultural belt concept proposed by Howard included both agricultural, recreational, and institutional functions, it was conceived of as a permanent thick boundary around the town. The town was a self-supporting satellite, part of a polynucleic system of centers around a larger city. The concept has since been modified, and made more flexible, as the "parkbelt" developed under the guidance of Raymond Unwin.[43] The "green girdle" was also promoted by Unwin, as a "possibly discontinuous chain of open spaces at the extremity of large cities."[44] A more practical model it was adopted by London in the 1930s. Since the Greater London Plan of 1944, Britain has been the most determined in creating and maintaining greenbelts, where they have protected agricultural land, controlled sprawl, separated cities, and provided recreation amenities.[45] Although they have not controlled decentralization, and, as Freestone notes, they have many other negative impacts including: increased land and house prices, increased car travel, and diverting development deeper into the countryside.[46] Other green space typologies developed from the urban park and garden city traditions of the nineteenth and early twentieth centuries. The "parkway," or "greenweb," is an American green space type that is created within the fabric of the city to create separations and corridors; first developed by Olmstead in Boston with the "emerald necklace" set of parks; an integrated system was developed in Seattle in 1903 and in Chicago in 1909.[47] The "green backcloth" uses the satellite city concept, set in large green areas around a major center. Promoted by Thomas Adams and others, is the "green wedge or corridor," exemplified by Copenhagen's "Finger Plan" of the 1940s, this typology is a important method for injecting large green space systems into the full structure of a city, and providing good linkages between the city and surrounding country. The "greenway" is

---

[41] Ibid., pp. 82-83.
[42] Purdom, *The Building of Satellite Towns*, p. 442.
[43] Freestone, "Greenbelts in City and Regional Planning," pp. 71-73.
[44] Ibid., p. 74.
[45] Ibid., p. 81.
[46] Ibid., pp. 82-83.
[47] Ibid., pp. 75-76.

often a community generated linear system that preserves ecosystems and can incorporate trails.[48] The "green zone" is a large permanently held green area that can be used as an urban growth boundary, and can include wetlands, golf courses, national parks, and conservation areas. There are many overlapping types of space, as Freestone states, citing Peter Calthorpe:

> These basically comprise a threefold hierarchy giving form and shape to the region: greenbelts that form a natural and ultimate edge to the urban field, open spaces that form a large-scale connecting greenweb within the region, and spaces that provide neighbourhood identity and recreation.[49]

Beyond the two original Garden Cities of Letchworth (begun in 1903) and Welwyn (begun in 1919 under Howard's leadership), agricultural belts transformed into greenbelts which tended to concentrate on preserving surrounding natural and agricultural areas. This tends to be the case in the three American towns of Greenbelt (Maryland), Greendale (Wisconsin), and Greenhills (Ohio) begun during the New Deal era using Garden City concepts although the results were mixed; notably a greenbelt was not used in the earlier Radburn (New Jersey) experiment.[50] Greenbelts have become a common planning mechanism in Britain to control growth and preserve landscapes around major metropolitan areas, although they were not adopted by the New Towns program begun in 1946. One of the most enduring examples is the Ottawa greenbelt proposed by Jacques Gréber in 1950 and since its implementation in 1956 managed by the National Capital Commission; a study undertaken in 1972 comprehensively documented the system at that time.[51] In recent years there has been much interest in the sustainable city, and linkages have been made to the Garden City.[52] Further, the American New Urbanist movement employs the greenbelt, and related typologies, on occasion.[53]

CONCLUSION: ACTIVE ECOLOGIES

The city and the country reflect differences in the intensity and density of occupation and function; in the country there is a greater distance between people.[54] The common connection between farming, gardening, and the concept for the Garden City is a unified notion regarding the productive role of labor, and a concerted effort to

---

[48] Ibid., p. 88.
[49] Ibid., p. 94.
[50] See C.S. Stein, *Toward New Towns for America* (Cambridge, Mass.: MIT Press, 1966); Joseph L. Arnold, *The New Deal in the Suburbs: A History of the Grenbelt Town Program 1935-1954* (Columbus: Ohio State University Press, 1971).
[51] See Don Page, *The Greenbelt* (Ottawa: National Capital Commission, 1972).
[52] See Peter Hall and Colin Ward, *Sociable Cities: The Legacy of Ebenezer Howard* (Chichester: John Wiley & Sons Ltd., 1988).
[53] See Andres Duany and Elizabeth Plater-Zyberk, *Towns and Town-making Principles* (New York: Rizzoli, 1991).
[54] See A. Ballantyne and G. Ince, "Rural and Urban milieux," in A. Ballantyne, ed., *Rural and Urban: Architecture Between Two Cultures* (Abingdon: Routledge, 2010), p. 14.

overcome the divide between city and country. Berry argues that contemporary society, driven by large economic interests, is "based on a series of radical disconnections between body and soul, husband and wife, marriage and community, community and earth."[55] He suggests that farming as a form of work, feeds the body in many healthy ways. Elsewhere, Berry discusses what he describes as "good" work, or work that is the "enactment of connections."[56] There is no doubt a traditionalist, or nostalgic, aspect to Berry's critique, however, he has exposed some of the deeper operational problems in the urban-rural system. Berry raises a number of key points in his description, particularly with regard to the complex knowledge that farming requires, knowledge that is passed down through farming cultures. However, there are also some conventional ideas about farming that Berry propagates that can be challenged. Firstly, the famous independence of the farmer can be questioned, as farming has always operated within complex economic, social, and ecological systems. The interdependence between cities and farms is an example, in that cities are the primary markets for produce, and cities tend to be where the innovations in farming practice occur. Secondly, Berry alludes to the constant struggle that farming (and gardening) entails, the battle against markets, pests, weather, etc. The militaristic references often employed in writings on gardening and farming seem to be worn out metaphors. The notion that gardening and farming are confrontational activities, involving continuous battle or struggle, negates the historic role of these activities. As writers like Michael Pollan and Berry underscore, when practiced in an ethical manner, gardening and farming are inherently ecological activities, resulting in managed and complex landscapes. While farming has always been an integral aspect of urbanization, until the emergence of the Garden City, gardening played a minor role and occupied a small territory; the Garden City attempted to radicalize the role of gardening and the spatiality of the garden.

---

[55] Berry, *The Unsettling of America*, p. 137.
[56] Ibid., p. 139.

CHAPTER 9

Letchworth Garden City Innovations

The Garden City drew together many ideas and forces that were operating in British society at the end of the nineteenth century. Factors as wide-ranging as labor improvement, land reform, housing design, gardening, agricultural practices, community development, infrastructure planning, site design, governance, development and financing, and technology. Many aspects of twentieth century urban planning were conceived or refined at Letchworth, many ideas that were both developed and rejected in subsequent iterations of Garden City thinking.[1] The first decade at Letchworth was an experiment in both the social and structural organization of a small town, some of the specific concepts employed at Letchworth include: the concept of the agricultural belt or greenbelt, architectural controls, the emphasis on relatively low density housing with gardens, the attempt to use gardening as a civic activity, the emphasis on improving the working conditions of agricultural and industrial workers, social and community planning, the development of distinct neighborhoods, and the zoning of functions. Developed during a time of social, political, and economic upheaval the Garden City produced new territorial arrangements and attempted new forms of urban agency. Despite the many social and political reforms contained in the model, it also represents a modern, and scientific approach to urban development.

The standard criticism of the Garden City movement, and the developments that would follow, is that there is no urban vitality in Garden Cities, and that like all small communities, social life is provincially dull and amateurish. In other words, there is no real intensity, or creativity, such as one would find in a large and dense city. The typical critique of the Garden City has also revolved around its low-density settlement patterns, which is seen as wasteful and inefficient. The Garden City is also directly linked to the explosion of urban sprawl after World War II, which remains controversial as the original model was intended to counter the uncontrolled expansion of cities. The Garden City stands in opposition to the contemporary focus on high density compact cities and traditional notions of the public realm. There is no doubt that the Garden City places emphasis on the commodious home and garden, over vital public space. The critique emerged as early as 1913 when A.T. Edwards

---

[1] See, for example, S. Buder, *Visionaries and Planners: The Garden City Movement and the Modern Community* (New York: Oxford University Press, 1990); P. Hall and C. Ward, *Sociable Cities: The Legacy of Ebenezer Howard* (Chichester: John Wiley & Sons Ltd., 1988); G.E. Cherry, *The Evolution of British Town Planning* (Heath & Reach: Leonard Hill Books, 1974); P. Hall, *Cities of Tomorrow: An Intellectual History of Urban Planning and Design in the Twentieth Century* (Oxford: Basil Blackwell, 1988).

argued that the Garden City was "a revolt against civilization,"[2] and has been repeated many times since. And yet the ideals embodied in the model have been remarkably enduring. A community is, at any given time, an assemblage of bodies, structures, passions, actions, practices, expressions, and territories that results in some productive force. What does Letchworth innovate, particularly during the period 1903-1913?

THE MANAGEMENT OF FLOWS

As discussed above, flows involve the interactions of fluids and the environments they encounter, also includes the movement of materials that approximate fluids. Flows carry many elements necessary for the functioning of a city or town, including energy, nutrients, organisms, information, and waste. Historically, cities have built infrastructure to handle flows, systems such as roads, water distribution, and waste removal, they also made adjustments to accommodate water, wind, and sun. In the nineteenth century there was a great expansion of transportation systems (including the railway), the birth of information technologies, and developments in water and waste management systems; all of these would impact the city dramatically. The development of Letchworth Garden City incorporated the latest in these flow management technologies.

A major reason for the Letchworth site was that it was bisected by the railway (the Cambridge branch of the Great Northern Railway); the rail provided an essential linkage to London, but also to numerous other locations in the region. The original station was temporary, but a permanent station was built in 1912. The rail link allowed for the movement of goods, people, and the like to and from Letchworth. However, the location of Letchworth would prove, over time, to be something of a hindrance as the town was not on a main rail line. The rail also provided a primary linkage to other towns in the region, this was the basis of Howard's Social Cities scheme which called for clusters of small towns operating together. The town originally accommodated buses and horse drawn vehicles and would quickly adapt to the onset of the automobile.

Of particular importance was the layout of a road system that needed to be connected to regional and national transportation systems, such as the Great North Road just to the east of Letchworth. It was also important as Unwin noted in 1913, to develop a functional system of streets for internal communication (consistent with the rapid development of the automobile), and to establish points for crossing under the railway.[3] With regard to the study of traffic flows in urban environments, Raymond Unwin, discussing the surveys of towns, wrote in 1909, "there should also be a careful survey made of general traffic; statistics should be prepared of its distribution and of

---

[2] A.T. Edwards, "A Criticism of the Garden City Movement," *The Town Planning Review*, Vol. 4, No. 2 (Jul. 1913), p. 157. See also T. Sharp, *Town and Countryside: Some Aspects of Urban and Rural Development* (London: Oxford University Press, 1937).
[3] See R. Unwin, "The Planning of the Garden City," in C.B. Purdom, *The Garden City: A Study in the Development of a Modern Town* (London: J.M. Dent & Sons, 1913).

the relative intensity from different districts of the daily inward and outward flow of population."[4] Further, he wrote regarding traffic,

> For the roads in a town to satisfy properly their primary function of highways, they must also be designed as to provide generally for easy access from any point in the town to any other. But they should provide, in addition, special facilities for the ebb and flow of particular tides of traffic, such as that from the outskirts to the centre and back again which daily takes place in most large cities, or that across the town from a residential district to a quarter occupied by works, factories, or other places of employment, or to important railway stations, harbours, and other centres of industry.[5]

Clearly, by 1909, and likely earlier, Unwin understood that the daily movement of people in cities as a flow system, in part, this would have been learned from his experiences developing Letchworth. The "special facilities" that he referred to, addressed the overall design of the town and transportation networks. He argued against the "trellis" or grid system, in favor of an "irregular radiating system," or spider-web form, that allowed for the effective subdivision of land and the creation of picturesque street views, along with inter-connection.[6] Unwin suggested that collision points and difficult junctions must be avoided so as not to impede flows of traffic, in other words, avoiding inefficiencies, or turbulence, in the system.[7] He commented on different modes and speeds of traffic, and ways of accommodating these in cities, referring to continental European examples. Further, he recognized that efficient road systems may be in conflict with the organization of building lots, and beautiful townscape design.[8] In some of his intersection diagrams Unwin included rudimentary streamlines that show the intended flow of traffic.

In his 1913 account of the engineering work for Letchworth, A.W.E. Bullmore (Engineer to the First Garden City Ltd.) provided an overview of the various systems. He noted that due to the fact it was a new town, the installation of "modern" services grew as the town grew.[9] He wrote that by 1913 "roads, sewers and sewage farms, waterworks and mains, gas works and mains, electrical works and cables, and railway sidings have been constructed."[10] He stressed the importance of a good topographical survey and knowledge of the geology in order to design an efficient and cost-effective infrastructure system. The initial water supply for Letchworth was secured in 1904 by boring into the chalk substrate and building a pumping station and reservoir; a second plant was built in 1907. By 1912 the annual water consumption for Letchworth was 85 million gallons; the establishment of a suitable water supply was followed by the

---

[4] R. Unwin, *Town Planning in Practice: An Introduction to the Art of Designing Cities and Suburbs* (New York: Benjamin Blom, Inc., 1971), p. 144.
[5] Ibid., p. 235.
[6] Ibid., p. 235.
[7] Ibid., p. 237.
[8] Ibid., pp. 240-248.
[9] A.W.E. Bullmore, "The Public Services of Garden City," in Purdom, *The Garden City*, p. 242.
[10] Ibid., pp. 244-245.

laying out of water mains throughout the town. The company originally decided to install gas for lighting, heating, and power for Letchworth. To this end a gas works was built in 1905, along with the requisite gas mains, the requirements for gas grew rapidly leading to several expansions of the gas works. In 1907 a power station for the generation of electricity was built, to supply the factories and center of town. Sewers were also integrated into the plan, along with permanent sewage farms for the disposal of waste; in his original text, Howard proposed that sewage and waste would be absorbed by the agricultural belt.[11] All of these systems handled the flows associated with modern cities, often segregating them from natural flow systems.

Related to the design of infrastructure, and the management of complex urban flows, was the health of the citizens of the Garden City, this included reducing mortality rates and the incidences of disease, a widespread concern at the time. Since the 1840s great advances had been made in all aspects of civil engineering, which had greatly improved water supply, sewer systems, power supply, etc.; this would lead to improvements in the health of populations and the lowering of mortality rates. In this regard, Howard was inspired by the work of the British sanitarian Benjamin W. Richardson, and his vision for the sanitary city of Hygeia. Drawing from nineteenth century science (and medicine) and the work of figures such as Edwin Chadwick, Richardson, in his essay "Modern Sanitary Science, a City of Health" argued for the improved circulation of air (ventilation) and water in order to keep living environments clean. In the essay, Richardson described an ideal healthy city for 100,000 citizens; important institutions include the hospital and cemetery, which he described in detail, and various modern infrastructure systems, he wrote:

> At a distance from the town are the sanitary works, the sewage pumping works, the water and gas works, the slaughter-houses and the public laboratories. The sewage, which is brought from the town partly by its own flow and partly by pumping apparatus, is conveyed away to well-drained sewage farms belonging to the city, but at a distance from it.[12]

Richardson proposed that all the systems would be carefully designed and maintained (and regularly tested) under one principal "sanitary officer," or "a duly qualified medical man elected by the Municipal Council, whose sole duty it is to watch over the sanitary welfare of the place."[13] Richardson's city would effectively eliminate all diseases. This built on the work of Chadwick who worked tirelessly through the middle of the nineteenth century to reform the poor laws and to develop sanitation systems. As an example, the Public Health Act of 1875 was a vital factor in improving living conditions in Britain.[14]

Purdom noted that statistically, by 1913, Letchworth had a significantly lower death rate than the rest of the country. He acknowledged the important reforms in

---

[11] Howard, *To-Morrow: A Peaceful Path to Real Reform* (London: Routledge, 2003), p. 27.
[12] B.W. Richardson, "Modern Sanitary Science, a City of Health," *Van Nostrand's Eclectic Engineering Magazine*, 14 (January 1876), pp. 21-42.
[13] Ibid.
[14] See S.E. Finer, *The Life and Times of Sir Edwin Chadwick* (London: Methuen & Co. Ltd., 1952).

sanitation, particularly water supply and drainage, instigated by Chadwick and others in the nineteenth century. He also argued that much of this reform did not include housing reform, a crucial aspect of the development of Letchworth.[15] Purdom wrote:

> The Garden City carries on the improvement in sanitary conditions by improving town life. Fogs, smoke, slums, crowded areas, and bad houses are not found here. Its clean atmosphere and rural surroundings make town life as healthy as country life. In the Garden City factories, the workers pursue their vocations under conditions as perfect as can be devised by modern industry, and their homes are not in dark courts or distant and dismal suburbs, but in pleasant roads and surrounded by plenty of space. The principles observed in the development of the town, the restriction of its size, the limitation of the number of houses to each acre, and the regulations as to the size of rooms and gardens, are the means by which the Garden City brings back health to the town. It abolishes daily travelling for the worker, it increases his leisure, and it gives him an opportunity to enjoy open-air pursuits.[16]

Health was a key factor in the design of Letchworth. The integration of systems, good housing, green spaces, recreation, governance, commerce, and employment produced a total environment, foreshadowing the planning and execution of towns and suburbs throughout the twentieth century.

At Letchworth, by controlling urban flows, supply and waste could be managed, resulting in efficient transportation, water supply, power supply, and waste removal systems. Any turbulence, or inefficiency, was eliminated as far as possible. Citing Richardson and others, Schultz and McShane write, regarding nineteenth century engineering: "Engineers, accustomed to thinking about unified systems, joined sanitarians in viewing the city as an ecosystem, a vast, integrated unit with the efficient functioning of one part dependent on the efficient functioning of all the parts."[17] At Letchworth the latest in urban infrastructure was incorporated into the design and development of the town. And while many of these systems remove flows from larger ecosystems, and strove to eliminate inefficiency, there was, as Schultz and McShane state, a rudimentary understanding of the city as an ecology. Many of the infrastructure systems and processes widely employed today in urban environments were developed in the latter half of the nineteenth century, and these were understood as flow systems.

---

[15] See Purdom, *The Garden City*, pp. 176-178.
[16] Ibid., p. 179.
[17] S.K. Schultz and C. McShane, "To Engineer the Metropolis: Sewers, Sanitation, and City Planning in Late-Nineteenth-Century America," *The Journal of American History*, Vol. 65, No. 2 (Sept. 1978), p. 403.

The Functional Organization of Land

The early Garden City is associated with the development of functional zoning, implemented through the design of Letchworth, which carefully organized the location of various uses (residential, industrial, commercial, institutional, etc.); these were then managed under the rent system that was employed. This approach controlled land use in a way that was intended to provide a rational method for organizing human settlements and preserving the health of citizens. The notion of the regulation of city property began in France in the early nineteenth century and became popular in Germany in the mid-nineteenth century. However, the advent of modern urban zoning occurred with the passing of the Town Planning Act in England in 1909, shortly after the establishment of Letchworth. By the early 1930s the United States and Germany had embraced the concept of urban zoning to the highest degree.[18] As Adams wrote in 1932, the concept of zoning is,

> the regulation by law of the uses of land (or) buildings, and of the height and density of buildings, in specific areas for the purpose of securing health, safety, convenience and general welfare. In practice the term "general welfare" includes regard for rights of property and economic considerations and may also include aesthetic considerations when the public is prepared to support these as essential to its welfare.[19]

According to Adams, zoning promotes the following "special values": 1) "co-operation between local authorities and property owners," 2) preventing "defective forms," "excessive densities," and "wrong uses of land," 3) a basis for developing infrastructure, and 4) managing the quality of architectural design.[20] The implementation of zoning, then leads to questions of how to address "non-conforming uses" and "relaxations." Adams identified both "rigid" and "elastic" forms of zoning regimes. He suggested that North America has followed more rigid functional zoning of cities that tends to eliminate the historic blending of urban functions, whereas Britain, for example, evolved more flexible approaches, which has allowed for supplementary zoning measures.[21] Adams recognized that elasticity in zoning had advantages, but he also noted that it is "particularly important that the boundaries of zones be fixed so as to prevent one area being developed to the injury of, or out of harmony with, an adjacent area."[22] This implies that urban boundaries are intended to be fixed and impermeable, a factor that tends to operate against elastic or flexible urban development. For example, a lack of elasticity in zoning practices is evident in "blighted" areas of cities, where zones that have become functionally or economically obsolete, often resulting in gaps in an urban fabric; this could be countered by more adaptive approaches to zoning change. As argued above, cities as artificial ecologies,

---

[18] Thomas Adams, *Recent Advances in Town Planning* (London: J. & A. Churchill, 1932), pp. 184-186.
[19] Ibid., p. 185.
[20] Ibid., p. 202.
[21] Ibid., pp. 188-191.
[22] Ibid., p. 191.

require elasticity, both in the administration of regulations and in the structure of the land itself, however, this is often difficult to achieve in the face of land ownership issues and political challenges. Unfortunately, the "scientific" approach to planning that was developed at Letchworth, and subsequently expanded upon with the establishment of the Garden Cities and Town Planning Association in 1909, resulted in the systematic segregation of urban functions. The increasing complexity of urban infrastructure and the development of modern industrial functions resulted in the widespread adoption of more organized approaches to urban design and management. This created, in most cases, a rigid urban patchwork system, that is difficult to modify over time. The advent of modern zoning tends to create strong boundaries between urban functions, often resulting in urban zones that are mono-functional and disconnected.

At Letchworth the Parker and Unwin firm was supported by various engineers and employees of the company (reflecting the resources and expertise available), whereas, the implementation of the second Garden City at Welwyn, beginning in 1920, involved a large and organized team of specialists.[23] Inherent to the often makeshift processes used at Letchworth, was a strong underlying commitment to the control of the planning of the town, the design of infrastructure, and the total aesthetic environment from architecture to landscape design. However, this control, as F.J. Osborn stated, required a balanced approach:

> The value of well administered controls in protecting the amenities of a town cannot be doubted. But they have to be handled wisely and tactfully, as well as firmly, if they are to be effective without seeming to the controlled to be pettifogging or oppressive. What is objectionable and what is permissible is not a matter of estate economics only; public sentiment must be taken into account.[24]

The inherent danger in this approach to the design control of urban environments is that the end result will lack vitality. This control extended to limits on the kinds of businesses operated from private residences, and the creation and strict management of architectural design controls. The use of architectural controls at Letchworth was part of a comprehensive effort to orchestrate all aspects of the town's design. Enforcing signage, additions, and gardens would be an ongoing challenge in the Garden City, where overall control was sought. For example, Osborn argued that the "proper care" of gardens should be enforced by housing estates, rather than the town, and that in his experience only 5-8% of gardens suffered from neglect, typically the front garden as the public face of a residence was more important to upkeep than the back gardens.[25] Benevolo notes, the Garden City developed the concept of "townscape" design with its emphasis on the total public environment including

---

[23] F.J. Osborn, *Green-Belt Cities: The British Contribution* (London: Faber and Faber Ltd., 1946), pp. 70-71.
[24] Ibid., p. 97.
[25] Ibid., p. 102.

plantings, furniture, permitted uses, control of signage, noise control, etc.[26] Further, the Garden City extended this control to a model for a region, one that embraced agriculture. Manfredo Tafuri and Francesco Dal Co summarize that the Garden City achieved an approach that integrated across scales, they write:

> the idea of the garden city introduced by Ebenezer Howard was at one and the same time the ultimate utopia and one of the first scientific models of urban planning on a territorial scale. Together with it rose a regionalistic conception that began to reach beyond the limited urban ambit to the much broader domain of the region as physical, economic, and social reality, and this quite apart from all ideal models.[27]

The strongest element in this integrated conception was Howard's emphasis on the "ward," otherwise described as a "neighbourhood unit,"[28] "superblock," or "district." As one of the distinct innovations of the Garden City this is the mediating structure between the residence, or dwelling, and the city as a whole. Benevolo writes, regarding the neighborhood, or district, as a repeating element in the city, that:

> The greatest complication lay in the fact that the problem of the city was a problem of maximum efficiency: here the problem was to give the community everything it needed to satisfy its various requirements; the problem of the district, on the other hand—if not understood in a purely quantitative sense—was a problem of gradation; it was a question of isolating a suitably-sized unit within the city itself and of seeing what services and what activities should be provided on this scale and what others on the larger scale relevant to the whole city. From this point of view, it was irrelevant whether the district was made up of one-family houses or of intensively built multi-family blocks.[29]

However, the concept of the neighborhood, an important legacy of the Garden City, would also, to some extent, be its undoing. The concentration on well-planned residential neighborhoods by Parker, and especially Unwin, produced a planning unit that could be widely applied to town and suburban schemes; it had particular appeal to the middle classes.[30] This became the basis for suburban expansion, rather than the Garden City as a total concept.

If we examine one of the early neighborhoods, or districts, at Letchworth we can determine its arrangement and function. It is the district defined by Howard Park to the West, the railway to the North, the Factory sites and Pixmore Avenue to the East,

---

[26] L. Benevolo, *History of Modern Architecture*, Volume 1 (Cambridge, Mass.: MIT Press, 1971), p. 358.
[27] M. Tafuri and F. Dal Co, *Modern Architecture/1* (New York: Rizzoli, 1986), p. 29.
[28] See M. Miller, "The Origins of the Garden City Residential Neighborhood," in K.C. Parsons and D. Schulyer, eds., *From Garden City to Green City: The Legacy of Ebenezer Howard* (Baltimore: Johns Hopkins University Press, 2002), pp. 99-130.
[29] Benevolo, *History of Modern Architecture*, Volume I, p. 357.
[30] Buder, *Visionaries and Planners*, p. 88.

and Pixmore Way to the South, and includes the working-class cottage estates of Bird's Hill and Pixmore. By 1913 this neighborhood was largely built, and beyond the residential uses, included a school and the Letchworth Hospital. As a functionally zoned area, it is conveniently adjacent to the factory or industrial zone, the park, which provided amenities, and to the nearby central shopping district and recreation grounds. These, as Unwin noted in *Town Planning in Practice*, represented different approaches to organizing plots of land, and siting buildings. The Bird's Hill's section involved more irregular planning to accommodate a hill and views, whereas the Pixmore area was more regular.[31] In the chapter entitled "Of Plots and the Spacing and Placing of Buildings and Fences," Unwin demonstrated an integrated approach to site planning, transportation design, building placement, and the organization of interior spaces. This was designed to achieve a picturesque effect, with suitable street compositions and views from individual houses. Further, the total effort applied to the design of all aspects of the environment achieved a continuum between interior (including furnishings), building, garden, street, neighborhood, town, and country. Unwin would argue that "civic art must be the expression of the community,"[32] it must express the "common life and stimulate its inhabitants in their pursuit of the noble end."[33] Citing the medieval town as precedent, and following Sitte, Unwin became an international champion of the art of civic design. This involved the pursuit of a beauty based on an understanding of the mechanics of everyday life that could have a transformative effect on a community. This implies that a formal design processes, employing specific precedents and languages, can result in a design with rich communal potential. As noted by Benevolo, this integration across scales, was an important achievement of the Garden City, and was perfected by Parker and Unwin at Letchworth.

Accounts of the first decade at Letchworth stress the pioneering aspect and diversity of the community, and the many opportunities for social interaction to arise.[34] Osborn who lived in Letchworth and Welwyn Garden City, has provided insights about the nature of social interactions in early Letchworth. He noted that both Garden Cities would eventually become very similar to comparable communities elsewhere in England.[35] Writing about the early social life in the two Garden Cities, Osborn wrote:

> Common to the social life of both towns is the background of a decent home for virtually every family, and local employment for most. At all times and everywhere the main interests of settled people, unless thwarted by environmental conditions, evolve around their homes, and to a less but important extent around their daily occupations. The fact that on the whole people live within a few minutes' walk of their work adds to effective leisure

---

[31] Unwin, *Town Planning in Practice*, pp. 346-348.
[32] Ibid., p. 10.
[33] Ibid., p. 11.
[34] See, for example, anonymous testimonies in M. Addren, ed., *Letchworth Recollections* (Wakefield: Egon Publishers Ltd., 1995).
[35] Osborn, *Green-Belt Cities*, pp. 114-115.

time considerably. Community life expands outward from the home life and factory life. It follows that the amount of time the inhabitants devote to fortuitous assembly in the streets, restaurants and places of entertainment tends to be less than in a city where dwellings are cramped and unattractive and hold little emotional or cultural interest.[36]

Here, Osborn argued, like Unwin, that it is the home (and garden) that is at the center of communal life. He stated that even in Paris this is the case among the working classes who cannot afford the benefits of a great city. An active life that revolves around the family may lead to less intensity in the streets, but, as Osborn wrote, not necessarily less community activity. He also suggested that those who rely heavily on commercial entertainment for socialization are those that lack social encounters and community life. Instead he defended the notion of participating in local amateur sports and culture, against watching professional equivalents. This, of course, touches upon the common critique of the Garden City (as well as suburbia) from design elites that they are dull places to live; the model also reflects a focus on the home found in Northern European and Anglo-Saxon cultures.

Reading a development map of Letchworth from 1913, indicates that the town was by then partially built out according to the Parker and Unwin plan. The Estate Boundary is clearly labeled, as are the various agricultural uses on the edges of the development. The various zones (institutional, commercial, industrial, residential, and recreational) are evident. These are typically defined by location and are segregated by roads. A slightly earlier map, from 1910, shows more clearly, as part of the development strategy the areas designated for Cottage Sites, sites for Workmen's Cottages, Residential Sites, and Small Holdings, each targeted for a different socio-economic group. One significant criticism of the plan is the separation of residential neighborhoods based on wealth and occupation, so that the working classes were always displaced from the middle classes, despite the egalitarian theories espoused by the movement. Further, the Factory Sites are also clearly labeled as the town strove to attract manufacturers. The neighborhood units are clearly defined, incorporating a range of uses including common recreational spaces. The overall result is a coordinated set of land uses, or patches, organized to relate to one another dependent on the adjacency.

The use of zoning introduces a patchwork system determined by function and location, and typically defined by transportation corridors. The preponderance of green space throughout Letchworth including gardens, parks, and the agricultural belt, provides a layer that acts like a green matrix, and which provides an ecological framework for the community. Along with zoning, important innovations of the early Garden City included the careful attention to the design and construction of inexpensive and commodious workers' cottages, the neighborhood unit, the agricultural belt, the emphasis on gardens, and the effort to retain land in common ownership. The concept of the town as a garden was a conscious part of the design, reflected in the broad streets, planned industry, the comprehensive planting of street

---

[36] Ibid., pp. 117-118.

trees, hedges, private gardens, and other forms of vegetation and greenspace.[37] The industrial aspects of Letchworth were planned from the beginning, and included dedicated sites with good services and transportation, well-organized amenities, and high quality housing for the workers.[38] The healthiness of the town was an important factor, but over time the inadequate rail service and lack of social opportunities would be detrimental to Letchworth. Examining the land issue and the Garden City, Benevolo writes:

> If private speculation could be eliminated, the buildings could be set as far apart as needful and there could be open spaces everywhere; the stimulus to unlimited growth would also disappear and the size of the town could be suitably established so that the countryside was within walking distance.[39]

The concept of a bounded self-sufficient community, in concert with a region, implies a certain ecological balance. However, as Benevolo notes, Letchworth was viable, but never fully self-sufficient and economic pressures led to the reduction in the land devoted to the agricultural belt, nevertheless, this did protect the limits of the garden city.[40] The patchiness inherent to cities is reinforced in the early Garden City model, with its emphasis on defined functional zones that are bounded by corridor systems such as roads, and railway lines. However, the extensive use of green space, including gardens, street plantings, parks, and the agricultural belt, would soften the division between adjoining patches. In other words, there is an elastic potential contained within the low-density fabric of the early Garden City, that in theory would allow for transformation over time through the modification of land functions and boundary changes. This is difficult to achieve within modern land tenure regimes and may have been easier to implement under the common land ownership model originally envisioned by Howard, nevertheless, a reimagining of cities must be a continuous process.

BOUNDARIES AND CORRIDORS

The history of cities has involved the creation of strong edges, or boundaries. These include the traditional separation of country and city as defined by fortifications and gates, and the subdivision of land. The emergence of the Garden City model can be conceived of as a series of attempted spatial reorganizations, in which traditional boundaries were both weakened and strengthened. This is reflected in the use of zoning methods, property systems, infrastructure, management systems, and the implementation of the agricultural belt. Firstly, the active zoning of the town into areas designated for largely mono-functional purposes (residential, industrial,

---

[37] C.B. Purdom, *The Building of Satellite Towns: A Contribution to the Study of Town Development and Regional Planning* (London: J.M. Dent & Sons Ltd., 1949), pp. 96-97.
[38] Ibid., p. 114-115.
[39] Benevolo, *History of Modern Architecture*, Volume I, p. 351.
[40] Ibid., p. 355.

commercial, recreational, etc.) required the implementation of internal boundaries. However, the widespread use of greenery, tended to soften the inherent boundary structures internal to the town. Purdom captured this aspect, in a description of the open spaces of Letchworth and their inter-relationships:

> In the Garden City the characteristics of the open space belong to the town as a whole. When you walk about the town for the first time what you notice most is the spaciousness of the streets and the great width of the sky. The roadways are, if anything, rather narrower than in other towns, but the trees and greenswards and the distance between the houses make the streets a continuous open space, to which occasional shrubberies and beds of flowers give additional variety. Each street has a slightly different character, so that you may walk around the town and think yourself to be in a garden all the while. In every part of it you can hear, if you cannot see, the unmistakable signs of the open country. In the very heart of the town the pleasant noises of the farm are heard. There is no need for formal public gardens where this experience is shared by all there is no occasion to mention parks and open spaces, for the mere idea of them never occurs to you.[41]

What is striking about this passage is the emphasis on spatial openness, the presence of greenery, and the proximity of the surrounding farms. Rather than relying on the urban park as a localized phenomena, green space was conceived as being generalized and open, the town is conceptualized as a garden. The use of formal park space was largely abandoned, a notion aided by the relatively small size of the community. In the Garden City a tapestry of gardens was created, a continuous carpet of greenery joined to the surrounding agricultural belt. The continuity of green space systems and vegetation resulted in a green matrix that was layered over infrastructure systems and was subject to the flows provided by climate and weather; this matrix tends towards creating a boundaryless condition. The extensive green systems inherent to the Garden City model, including the comprehensive planting of trees, did mitigate against the bounded structures of the city, and the engineered infrastructure systems; for example, continuous tree canopies effect heating and cooling in cities, and manage air pollutants, further, they influence air flows, and create habitats.

Secondly, the subdivision of land further strengthened the bounded conditions throughout the town design. Within the neighborhood units, were further levels of boundary, that separated plots of land from one another. Recognizing that the English like to delineate their gardens (and properties) with fences, Unwin suggested that there is a need for "welding the plan of the site, the buildings, and the gardens, more and more into a complete whole."[42] This implies a kind of urban ecology, he further wrote:

---

[41] Purdom, *The Garden City*, pp. 112-13.
[42] Unwin, *Town Planning in Practice*, p. 381.

> With a co-operative society it is safe to count also on the common enjoyment of much of the garden space. It is, indeed, possible, even where houses are sold to individuals, to arrange some degree of associated use of gardens there may be introduced into our town suburbs and villages that sense of being the outward expression of an orderly community of people, having intimate relations with one another, which undoubtedly is given in old English villages.[43]

While the Garden City company originally held land in common, leasing plots to individuals and businesses, the notion of private property still pervaded the organization of Letchworth.

Thirdly, the infrastructure systems employed to manage flows, including transportation, water supply, sewers, and power supply, tended to reinforce boundary systems. The transportation system, where conduits function also as boundaries, typically has a significant impact on the organization and functioning of land within a town or city. At Letchworth, of particular note was the railway that subdivided the town, but also provided a vital linkage; difficult to cross, it acted mainly as a conduit, although the right-of-way could also provide some habitat. As is typical the roadway network in the town, carefully designed aesthetically, also subdivided the town into zoned areas, and neighborhoods. This created further levels of division within the structure of the town, and established differentiation between functions, and social classes.

Fourthly, and most significantly, the use of the "agricultural belt" (or greenbelt) attempted to reorganize the traditional relationship between farmer and urbanite, and between city and country. The separation between city and country is addressed in a unified scheme, through the creation of a thick and fluctuating boundary condition designed to negotiate between the town and the surrounding agricultural land in a productive way. The agricultural belt also redefined the historic urban boundary or wall, as it mediates between the town and the surrounding agricultural land in a productive way. Describing the agricultural belt, Purdom wrote:

> This rural belt surrounds the town like the walls of a mediaeval city. It limits its boundaries, protects it from the attack of other towns, and preserves its shape and style. It also gives it finish and completeness. The ragged edges of the ordinary town are not found in Garden City. There, where the town finishes the country begins.[44]

The agricultural belt was intended to be a boundary, to provide open space, to separate from adjacent communities, and to be used for agricultural purposes.[45] It was designed to support institutions, natural preserves, and rent paying farmers, typically small holdings. The Garden City, as first manifested at Letchworth, was characterized

---

[43] Ibid., pp. 381-382.
[44] Purdom, *The Building of Satellite Towns*, pp. 115-16.
[45] Ibid., p. 442.

by what Purdom described as "open space," a spatial system that is a refuge from the "restrictions of the town."[46] He argued that this was a comprehensive spatial idea that pervaded the town. Along with pervasive greenery, two key aspects to the spaciousness of the town were Norton Common, seventy acres of woodland, and the agricultural belt around the community; two thirds of the land assembled for Letchworth was maintained as farms, orchards, parks, and small holdings. Supported by the compact size of the community, these features contributed to a continuity between town and country.

The agricultural belt, as a concept for uniting town and country, was an attempt to revolutionize the fundamental relationship between the two. As noted above, one of the major concepts for the development of the agricultural belt was to limit growth, and to eliminate the land speculation found on the edges of most towns. With a pre-determined population limit, Howard's scheme involved the purchase of inexpensive agricultural land, and then through the common ownership of the town and the charging of rent based on land only, allowing the citizens of the town to benefit directly from the increased value placed on the land by urbanization. The controlled improvement of land was to produce dividends for the company, what Howard termed the "unearned increment."[47] The agricultural belt ensured that the land value for surrounding farms would remain relatively stable. Further, the agricultural belt was intended to redefine the farm and its relationship to the city. By developing a local market for farm products, the surrounding farms could be intensified. The historic dichotomy between the farmer and the urbanite or town-dweller (the manufacturer) was to be overcome. The growing of food at Letchworth occurred on farms, small holdings, and allotments within the agricultural belt, and in small private gardens adjacent to houses. According to H. Burr, Surveyor to the First Garden City, Ltd., writing in 1913, there were approximately 300 acres used for small holdings of 0.25 acres (or larger), and 12 acres held for allotment gardens.[48] By 1949 the area devoted to allotments had grown to 71.5 acres,[49] here the roles of the farmer and the gardener overlap.

The agricultural belt has also become an ambiguous kind of space encircling many cities around the world. But as the concept for the greenbelt developed it produced various problems, including increasing land costs, and dispersing development beyond the belt.[50] In the original vision of Howard's the agricultural belt played an active role in the life of the town, providing an interim layer between town and country. However, many subsequent greenbelts, while they conserve and preserve land, do not necessarily play an active role. Unlike a shantytown, or a patchwork system, it can be difficult to activate greenbelts within larger ecological systems; they tend not to function as a smooth space instigator.

---

[46] Purdom, *The Garden City*, p. 112.
[47] Ibid., p. 26.
[48] Osborn, *Green-Belt Cities*, pp. 272-83.
[49] Purdom, *The Building of Satellite Towns*, p. 145.
[50] See R. Freestone, "Greenbelts in City and Regional Planning," in Parsons and Schulyer, eds., *From Garden City to Green City*, pp. 82-83.

The Garden City provided an urban model where the agricultural or rural belt was integral to the plan of the town, therefore, farming and the farmer were also joined. The stated aim of the Garden City movement was to unite city and country,[51] to draw from the best of the two worlds, and to break down what was seen as an artificial divide; in other words, to activate the historic boundary between town and country, allowing for a greater range of interflows. Supporting the underlying contention of the Garden City movement, the British literary theorist Raymond Williams, in his text *The Country and the City*, tends to underscore the historic divide between urban and rural populations and territories. However, he does emphasize that while it has a long history, much of this division is a result of capitalism and industrialization.[52] The use of the agricultural belt, or greenbelt, as a wide boundary separating the town from the country and providing a protected zone for agricultural and cultural amenities, helped create the sense of a town engulfed in greenery. This is augmented by other green spaces in the town itself and enhanced by the generally single-family housing fabric and the emphasis placed on private gardens. As a satellite to a larger city, although intended to be self-supporting, the Garden City was, in part, designed to provide a familiar environment to those migrating to the city from the country.

It is evident that the Garden City model developed at Letchworth propagates many of the structural features historically associated with cities. As noted above, boundaries function as filters, conduits, habitats, sources, and sinks. The filtering effect created by the range of boundary systems at Letchworth, attempted to control flows between town and country, and between functions. The boundary also acting as a conduit is evident throughout the town. The notion of a boundary providing a habitat, source, and sink applies most specifically to the agricultural belt at the large scale, and hedgerows between gardens at the small scale. The creation of boundaries, and defined patches of land, to regulate flows is inherent to striation of space associated with the urbanized peoples. The ability to move towards a smooth space arrangement is most evident in the agricultural belt, and the pervasive employment of green space and plantings. These both have the potential to modulate flows outside of channelized infrastructure systems, and to accommodate changing conditions (seasonal, functional, etc.). The Letchworth plan employs a system of centers (territories) and peripheries (boundaries) throughout.[53] As noted above, these tend to be highly static and are determined by function, land ownership, class, and the like. The ability for patches (territories) and boundaries to be modified is limited. Within any ecology, constructed or otherwise, there will be boundaries of one kind or another, it is not possible to achieve a purely smooth space except in the most expansive of conditions, such as the sea, desert, or steppe. This suggests that in urban environments, there are boundaries that can have a positive effect on city and the various historic edges of cities (fortification walls, greenbelts, uncontrolled edges, etc.) each have their own effective influence on surrounding territories. Ensuring the

---

[51] See "The Three Magnets" diagram published in Howard, *Garden Cities of To-morrow*, p. 46.
[52] Raymond Williams, *The Country and the City* (Frogmore, St Albans: Granada Publishing, 1975), p. 366.
[53] See S. Sassen, "A New Geography of Centers and Margins: Summary and Implications," in R.R. LeGates and F. Stout, *The City Reader* (London: Routledge, 1996).

integrated functioning of defined parcels of land within a city, means understanding the complexity of urban functions, and the flow systems vital to this functioning.

AGENTS AND AGENCIES

During the establishment of Letchworth a number of key agents were brought together motivated by the ideas embedded in Howard's text. This involved individuals on one hand, with a wide range of skills, including figures such as the gardener and farmer who were considered vital to the creation of the model. On the other hand, the establishment of a new town required setting up various administrative agencies to implement the overall plan. Cities have always employed complex administrative systems, these have tended to be hierarchical and authoritative, organized by departments with clear boundaries and specific tasks. Nevertheless, beyond individual and organizational actions, there is also the potential in urban structures for a system to self-organize, or to generate unexpected arrangements or productions. The definition of organizational systems, and the type of boundary mechanisms employed will determine how effective a bureaucracy will be in managing the operations of a town. As DeLanda notes this has to do with actual, and perceived, forms of jurisdiction[54]; jurisdiction has material (bureaucrats, buildings, etc.), expressive (legal codes, procedures, record keeping, etc.), and territorial components. These kinds of organizational structures can result in conflicts between jurisdictions, inefficiency, or inaction.

In his original text, Howard devoted much to the financing of the Garden City, to questions of land ownership and rent, and to the administration of the town, throughout it is evident that he was very suspicious of the role of the State in British society, as he went to great lengths to describe a community that was collectively held, this was to be achieved through the charging of land rents that would provide the financial means for building and maintaining the town. Howard suggested that "progressive" organizations could likely rise above common aspirations, and that it "is probable that the government of a community can never reach a higher tone or work on a higher plane than the average sense of that community demands and enforces."[55] He further argued that communities would be well served by those with a higher sense of "social duty."[56] Howard believed that Communism and Socialism held many important values, however, he also championed the role of the Individual, so that idiosyncratically, he advocated for a system that was strong collectively and supported the individual; this reflected Howard's strong pragmatism.[57] Therefore, as a

---

[54] M. DeLanda, *A New Philosophy of Society* (London: Continuum, 2006), pp. 73-74.
[55] Howard, *To-Morrow*, p. 82.
[56] Ibid., p. 82.
[57] Ibid., pp. 94-101. Deleuze and Guattari's concept of the assemblage is effectively apolitical, however, assemblages are often political in nature. Deleuze and Guattari, were typically considered to be leftist thinkers, however, Deleuze's political position in particular is complex. Unlike Howard, they did not advocate for the concept of the individual, believing more in collective assemblages. However, like Howard, Deleuze and Guattari attacked the concept of the State, however, they supported the concept of nomadic organizations, and localized (micropolitical) systems. Focusing on the concept of "power," in the context of

response to Howard's own political views and the land nationalization movement the Garden City attempted to demonstrate that the common ownership of land (through a commonly held Garden City company which is the only landlord), and rents based on land, would result in a more egalitarian society. Rent would be collected by trustees who would manage any debts, and who would work with the Board of Management to undertake municipal works, and the like.

Out of necessity, the development of the administrative structures for Letchworth was based on contemporary business models; the Board of Management would consist of a Central Council and various Departments (Public Control, Engineering, and Social Purposes). Howard believed his model, although conventional in commercial terms, would be better at "carrying out the will of the people."[58] Howard described a clear hierarchy in the functioning of the various entities (council, departments, and groups). Describing Diagram No. 5 that laid out the organization, he wrote: "At the centre is the Central Council with its ample powers of co-ordination and financial control, upon which body sit the chief officers of all the public departments."[59] The organization had a conventional zoning of groups by function, with boundaries defined for each. Purdom provided a succinct description of the company and its structure:

> The First Garden City, Ltd., which is the owner of the freehold of the land on which the town is built, as well as the organization which has developed the place and provided its public services, is a joint-stock company registered under the Companies Acts. It is governed by a memorandum of association, which is practically unalterable, and by articles of association, which may be altered at any time by its members and contain the regulations for the conduct of the company's affairs.[60]

All the financing for the venture was raised from private investors. There were chronic difficulties in raising capital, and limits were placed on returns on investment (dividends), which were intended to discourage speculation. Three committees (finance, building and letting, and engineering) managed the business of the company, these roughly equated to Howard's original structure. The directors of the company met with the consultative committee of the shareholders (which in 1913 numbered 2400) three times a year. Through this structure, and its staff, the company undertook the following operations:

Deleuze and Guattari's political position, it can be described as being "conceived as the basic interaction of pluralistic, local, differential and productive forces that provide much more diverse and heterogeneous effects than those that could be identified along the conflictual, striated lines of institutionalised social actors and conflicts." See R. Krause and M. Rölli, "Micropolitical Associations," in I. Buchanan and N. Thoburn, eds., *Deleuze and Politics* (Edinburgh: Edinburgh University Press, 2008), p. 241.
[58] Howard, *To-Morrow*, p. 68.
[59] Ibid., p. 91.
[60] Purdom, *The Garden City*, p. 183. The accounting for the first ten years of Letchworth is provided in Appendix C of Purdom's book.

In addition to being the ground landlord, the company provides and manages the gas, water, electrical, sewage disposal works, has its own gravel and sand pits, constructs its own roads and buildings, manages farms, a hotel, a public omnibus, and swimming bath, and maintains open spaces. It also passes plans and inspects all new buildings.[61]

Therefore, despite the many of the novel ideas contained in Howard's vision, the town, as a private enterprise undertaking, followed business and municipal conventions of the day. The major difference being that it did so outside of the typical governmental support that similarly sized towns in Britain would have enjoyed; the town was eventually nationalized in the early 1960s.[62] The years of struggle to secure adequate capital to finance the development, tended to compromise some of the key aspects of the scheme. Howard's focus on land rents was one of the innovations of the vision, however, his financial model was quickly abandoned after the establishment of Letchworth and replaced with more typical forms of rent management.

Beyond the formal governance of the Garden City, Howard proposed various other kinds of agency including: The Semi-Municipal Group which included such enterprises as markets, the Pro-Municipal Group that included educational, medical, and financial organizations, and the Co-operative Financial Group that included businesses, farms, clubs, factories, and homes. Howard noted the vital role of philanthropic, charitable, religious, educational, and other agencies in the development of a town. When it came to building houses Howard stressed the importance of lenders, particularly co-operative societies, against the practices of speculative builders. The development of Letchworth as a private enterprise undertaking, is testimony to Howard's distrust of government, to this end, the various companies were formed and managed by forward thinking industrialists, lawyers, members of Parliament, gentry, and the like. The various early figures that coalesced around Howard and the Garden City concept became the agents of the Garden City movement, a group very devoted to implementing the ideas set out in text.

However, Letchworth was not always well served by its administrative system, with key figures not completely committed to the development of Letchworth and its ideals, particularly Thomas Adams' successor W.H. Gaunt, who was the estate manager from 1906 to 1917. This resulted in many compromises. Further, according to Purdom, the Board of Directors was disconnected from the day-to-day operations of the town, often resulting in poor, or highly conventional, decisions being made.[63] Nevertheless, a number of individuals would be instrumental in establishing Letchworth. A vital figure in the implementation of Letchworth was Sir Ralph Neville (1848-1918), a prominent barrister, who was the first Chairman of the Garden City Pioneer Company, and the subsequent First Garden City, Ltd. It was he that turned many of Howard's ideas into a practical reality, including establishing a more

---

[61] H. Craske, "Appendix C: The Garden City Company," in Purdom, *The Garden City*, p. 241.
[62] See P. Hall and C. Ward, *Sociable Cities: The Legacy of Ebenezer Howard* (Chichester: John Wiley & Sons Ltd., 1988), pp. 29-46.
[63] C.B. Purdom, *Life Over Again* (London: J.M. Dent & Sons, 1951), p. 60.

conventional approach to leasing land from that outlined by Howard.[64] Various early members of the Board including Aneurin Williams, and the industrialist Edward Cadbury were also instrumental figures. Neville hired Adams, as the first full-time secretary, who also played a key role in both the early development of Letchworth, and subsequently in the dissemination of Garden City principles. Both Neville and Adams resigned from the initiative in 1906, which was a blow to the momentum of the project. Purdom, as noted above, was an early employee of the Garden City company, and would also be a lifelong advocate. Other important figures included Parker and Unwin, whose firm produced the plan, and many building designs for Letchworth. These men were instrumental in establishing the first Garden City, as Howard's role effectively became a symbolic one. During World War I a struggle for control of the Garden City movement occurred in which both Purdom and F.J. Osborn (1885-1978) were vital figures, this was the point that Osborn joined the enterprise; both would be lifelong promotes of the Garden City cause. Later on, there would be many other key promoters, including Lewis Mumford.

Also essential to Letchworth's viability were the many pioneering inhabitants who came to the town in the first decade, some attracted by its idealistic vision, and others out of necessity. This raises the question of the community as a complex set of assemblages, each with its own participants, organizational codes, and structures; do these play a productive role in a community, or merely propagate already established practices? The early pioneers at Letchworth brought with them ideals that influenced their modes of work, recreation, religion, and lifestyle, which led to the formation of societies, institutions, and organizations. Beyond the designers who realized Howard's ideas, were those who moved to Letchworth during the first years, pioneers who were often required to develop social systems and modes of practice. Notoriously, the original settlers who were involved in a frontier experiment, were often seen as radicals and eccentrics, however, as the town developed the population quickly normalized. During the first decade a hospital was built, churches and schools were established, businesses developed, and a wide range of cultural and recreational organizations emerged. Purdom noted that originally the principal arts of Letchworth were architecture and gardening, but these were quickly supplemented by other disciplines, particularly music, theater, and the industrial arts, as inspired by the Arts and Crafts movement.[65] The practice of the arts by "amateurs" was integral to the early life of the community, as was participation in sport. F.J. Osborn, who lived in Letchworth for five years, beginning in 1912, and for many years in Welwyn Garden City, has provided insights into the social and cultural life of the early Garden City. He noted that the population composition of the two towns was very similar to those in similarly sized towns in southern England.[66] He further noted that the first pioneering years in both Garden Cities the social life was very active, although challenged by a lack of amenities. He addressed the common criticism of Garden Cities (and New Towns) that they are socially and culturally dull, by arguing that

---

[64] See Buder, *Visionaries and Planners*, pp. 79-81.
[65] Purdom, *The Garden City*, p. 134.
[66] Osborn, *Green-Belt Cities*, pp. 114-116.

residents were nevertheless very active in a wide range of local activities from sports, cultural societies, political organizations, horticultural groups, etc.[67] Osborn suggested that living in a small town demands more responsibility from those who are engaged in public life, and that the Garden Cities were generally tolerant places, not burdened with provincialism and parochialism; he acknowledged that some cultural amenities were lacking in both towns.[68] Cities are managed by complex bureaucratic and political structures that provide systems of agency. As noted above, at Letchworth these tended to be relatively conventional, although they operated outside typical public systems of governance. The private nature of the enterprise was unique, although difficulties raising sufficient operating capital was endemic to the growth of Letchworth.

The layout of the town also invoked the role of various activities such as gardening, farming, and recreation; these provided an important communal life for a town relatively devoid of cultural opportunities. What Letchworth lacked in urban vitality it tried to make up for in other ways. However, the highly regulated and planned nature of Letchworth tended, as much as possible, to eliminate the unexpected, or the spontaneous. Systems were carefully arranged to function smoothly, as was the functionally clear organization of the town. For example, despite the efforts of the founders, the working and middle classes were largely segregated in space and time. Further, the emphasis placed on the private realm of the home at Letchworth, what Richard Sennett has called the "tyranny of intimacy,"[69] was essential to the model. The careful design of residential and adjoining garden environments was primary, somewhat paradoxically the private realm, through labor, was to create a kind of community. However, as a self-organizing system, or productive set of assemblages, there is no doubt that there occurred at Letchworth during its first decade, many social and communal opportunities; these ranged from church groups, recreational and cultural organizations, to political associations. Like Howard's text, the controlled infrastructure of the town, produced during Letchworth's first decade many experiments in community building.

With respect to the general question of agency, and finances at Letchworth, Hall and Ward write:

> We have seen that, ever since Howard's day, these interrelated questions have proved some of the most intractable: we [the British] have never been able to devise a solution that successfully combined public and private agencies and financing, and we have never, from Letchworth onward, come near Howard's vision of a self-governing, self-financing commonwealth.[70]

By avoiding, at least initially, the support of the British state, and operating as a private company, Letchworth tended to function as a commercial enterprise. The

---

[67] Ibid., pp. 118-119.
[68] Ibid., pp. 122-128.
[69] See Richard Sennett, *The Fall of Public Man* (New York: Vintage Books, 1978), pp. 337-340.
[70] Hall and Ward, *Sociable Cities*, p. 121.

enterprise began as a company, however, this led to inevitable conflicts between those trying to uphold the vision of the Garden City movement, and financial realities. The company directors were often charged as being "paternalistic" towards residents.[71] There were, no doubt, socialists and extremists among the population who conflicted with the business interests of both the company and some of the industries that moved to Letchworth; this was also reflected in attitudes displayed by some of the Garden City company staff towards more eccentric members of the community. As noted above, the Garden City emerged at a time when many social and political reforms were underway. Science was invoked in planning, and the professions, including planning, were becoming more authoritative and active.[72] The Town Planning and Housing Act, passed in 1909 by the British parliament, institutionalized modern town planning, and the Garden City movement was an integral part of this. This led to the formation of the Garden Cities and Town Planning Association in the same year. Buder argues that the various pressures put on the movement caused it to lose direction, he also cites a memorandum written by Adams in 1904, foreshadowing this, urging the movement to become broader in its approach.[73]

At Letchworth, the ideas of Howard, inspired the birth of a movement, which would attract many followers. Howard was not a particularly inspirational figure, however, his text, despite its pragmatic tone, galvanized action; by focusing on issues of land, financing, and the like, his "utopian" vision struck a chord in early twentieth century British society. Howard was able to perceptively draw together a diverse set of ideas into his model for the Garden City, leaving it open enough for interpretation and transformation. Howard shaped a vision, with a set of diagrams, plans, and codes that were translated into a constructed and social reality as an experiment in integrated living. The results at Letchworth were not always spectacular, nevertheless many of the concepts were enduring and would be applied both selectively and comprehensively across the world. Therefore, it may be claimed that Howard and his key supporters, were philosophers of the contemporary city, employing science, observation, and invention in their pursuit of higher standards of living for the working and middle classes. Harnessing philanthropic private enterprise, which was key to the establishment of Letchworth, turned out to have enormous impact on subsequent urban legislation. Balancing public and private enterprise and ensuring that the administrative model employed by the pioneers of early Letchworth was not too radical, inevitably led to the compromises noted above.

Agency is inherent to the concept of assemblage (in French *agencement*), in that assemblages employ agents, both active and inactive, and the entire machinic construct should have agency, in that it has emergent properties. Here we have looked at both the role of specific agents in the development of the Garden City, and the agency inherent to various aspects of the model. When it comes to determining the most innovative agencing aspects of the early Garden City model we can look to

---

[71] See S. Meacham, *Regaining Paradise: Englishness and the Early Garden City Movement* (New Haven: Yale University Press, 1999), pp. 124-125.
[72] Buder, *Visionaries and Planners*, p. 97.
[73] Ibid., p. 98.

Ebenezer Howard and his book, the role of the designers, and impact of the figure of the gardener. Ebenezer Howard by all accounts was a humble and rather unimpressive person, who was a good public speaker, but not well suited to the practicalities of developing a new town. Nevertheless, there is no denying the impact of his ideas, primarily through his famous book. As Deleuze and Guattari write, "an assemblage, in its multiplicity, necessarily acts on semiotic flows, material flows, and social flows simultaneously."[74] Ostensibly describing the writing of *A Thousand Plateaus*, they are also describing the role of a book as an assemblage. The neutrality in the writing and lack of detail in Howard's text allowed it to be widely interpreted, the ideas embodied in *To-Morrow: A Peaceful Path to Real Reform* were realized by the architects and the team of specialists assembled to undertake the design and early development of Letchworth. It was the vision developed by Parker and Unwin, despite their relative lack of experience that galvanized the imaginations of town designers and builders around the world.

Traditionally we associate various figures as occupying the public realm, and others occupying the private. In the contemporary city these distinctions have become blurred, in particular as new urban figures have emerged. There is no doubt that while public spaces and amenities were provided at Letchworth, the emphasis was placed on the private house and garden. Traditionally, as Richard Sennett points out, the streets and squares of the city represented all of society participating in a great public theatre, the "gathering of strangers," where everyone is both an actor and a spectator, performing roles based on their station in life. [75] Public space was a space of commerce, communication and movement. Prior to the twentieth century, urban parks were places of leisure where one could see and be seen, be involved in recreational activities, or merely enjoy nature.[76]

With respect to the concept of assemblage and agency, Letchworth activated the seeker of alternative forms of living. Figures such as the gardener, the industrial worker, and the agricultural worker, instead of the urbanite, were foregrounded. To varying degrees each of these figures are the agents of striated space located in the garden, in the factory, on the farm, or on the road. In an attempt to reduce the migration from rural areas to cities, Letchworth successfully revitalized agriculture in the local area, and improved the lives of agricultural laborers by providing better wages and social amenities.[77] Letchworth was also designed to improve the lives of industrial workers drawn to the community by either necessity or choice. Gardeners typically represent a rural, or pastoral, or even suburban, ideal; solitary figures toiling away in gardens or anonymously maintaining parks. And yet, the gardener and the garden can provide operational answers, as cities must increasingly face significant environmental challenges.

---

[74] Gilles Deleuze and Félix Guattari, *A Thousand Plateaus: Capitalism and Schizophrenia* (Minneapolis: University of Minnesota Press, 1987), pp. 22-23.
[75] See Sennett, *The Fall of Public Man*, pp. 3-44.
[76] Ibid., pp. 85-86.
[77] See Purdom, *The Building of Satellite Towns*, p. 146.

THE GARDENER AS A COMMUNITY FIGURE

When it comes to the role of specific agents in the establishment of Letchworth, the gardener figured as the most prominent, and potentially the most innovative. The gardener was fundamentally embedded in Parker and Unwin's plan, and implicit in Howard's text. The Garden City attempted to unite country and town through the garden, particularly the space of the private residential garden and the physical act of gardening. The socialist emphasis on meaningful manual labor, as outlined by thinkers such as Edward Carpenter and Peter Kropotkin, would be translated by the Garden City movement into an ideal for active and communal participation in gardening. It must be remembered that much of the effort of the Garden City movement was directed at improving working conditions for both the urban and rural poor, and that enlightened manufacturers were integrated into Letchworth from the beginning. The Garden City, consistent with nineteenth century socialist thinking, attempted to reverse the trends created by the Industrial Revolution. Therefore, beyond providing good working conditions in the various factories attracted to Letchworth, was the expectation that the entire community would enjoy the benefits of gardening. This was integral to the model, and its employment of quality housing and generous gardens attached to each residence.

What is particularly innovative about the Garden City model was that, based on a celebration of rural life, an urban model was proposed where gardening was worked into the structure of the development; this was consistent with a widespread interest in "amateur" gardening in great Britain during the nineteenth and twentieth centuries.[78] As reinforced by Unwin in his essay "Nothing Gained by Over-crowding!" this resulted in a low-density urban fabric with an emphasis on single-family homes each with its own front and rear gardens.[79] The lowering of density and the reconfiguring of the design of blocks resulted in larger private gardens, neighborhood green spaces, and a reduction in the amount of road surface. In Unwin's essay, there are two important diagrams that he provided: the first shows a comparison between a typical bylaw street pattern and the superblock pattern developed for Letchworth; the second compares the land allocated to gardens (including recreation ground), houses, and roadways in the two schemes. In the second scheme, as employed at Letchworth, there is a dramatic reduction in area devoted to houses and roadway, and a significant increase in space for gardens and community space. This clearly reflected the emphasis on gardening as a vital activity, and the figure of the gardener, in the plan of the Garden City.

At Letchworth Parker and Unwin took into account many factors in the design of the town, however, one of the strong guiding principles was the desire to achieve picturesque effects in the design of streets, plantings, and buildings. Unwin made the following statement regarding the interconnection between the individual and the community:

---

[78] See A. Wilkinson, *The Victorian Gardener: The Growth of Gardening & the Floral World* (Stroud: Sutton Publishing, 2006).
[79] See R. Unwin, "Nothing Gained by Over-Crowding!" in Walter L. Creese, ed., *The Legacy of Raymond Unwin: A Human Pattern for Planning* (Cambridge, Mass.: MIT Press, 1967), pp. 109-126.

> The gardener, like the architect, has fixed his eye too exclusively on the individual plot; he has thought too much of the bulbs in his own individual beds. We need to think of the street, the district, the town as larger wholes, and find a glorious function and a worthy guidance for the decorative treatment of each plot and each house in so designing them that they shall contribute to some total effect. For is it not a finer thing to be part of a great whole than to be merely a showy unit among a multitude of other units?[80]

Unwin equated the gardener with the architect and charged both with looking at the whole effect. While he tended to emphasize the visual and the picturesque, there is the implication of broader thinking, including the importance of site planning. In his chapter on this subject in his text, he argued for the careful survey of the site, and for the preservation of key features, including views and existing vegetation; he addressed in detail site and the orientation of buildings to the sun for best exposure. He also devoted much text to density, shape and size of plots, and the siting of buildings. He argued for the long narrow plot, as used at Letchworth, which allowed the garden to have ground "in the most valuable position; and having one long dimension, good vistas may be developed."[81] Unwin also advocated for the use of architectural controls to ensure both harmony of colors and materials, and to guide the overall composition of buildings for public effect. He concluded his book on town planning with a discussion of the strengths and weakness of by-laws. As an example of controlling environments, he noted that the design of fences and entrance gates must be submitted to the company for approval, and if adjacent landowners mutually agreed then they could choose not to have a fence or hedge.[82]

At the end of the nineteenth century there was not the ecological knowledge that we have today. Still, there was a belief that the city could be a garden, comprised of gardens, within a larger system of gardens (or farms). The gardener was put forward as an ideal urban figure by the Garden City movement, due to the vital labor involved. In the Garden City, gardening would provide a moral basis for a society, and reconcile the divide between country (agriculture) and the city. In his account of Letchworth's first decade, Purdom gave further insight into the role that gardening was intended to play in the Garden City community. Gardening was promoted as an activity that was meant to involve the majority of citizens, and to be an activity that had moral and physical benefit. Purdom wrote that a garden is good for "exercise, for sleeping in, for lovers, for children, for cats, for the spending of money, and gardening."[83] He recognized that the garden is a complex construct, able to support many activities and structures; a garden is ornamental and useful, a place for repose and for growing food, it is the place of both poets and the cultivator. At Letchworth every house had a garden and the occupants of the house were required to keep the garden in order.[84] If a garden was not properly maintained the Garden City company would put the garden

---

[80] Unwin, *Town Planning in Practice*, p. 288.
[81] Ibid., p. 350.
[82] See Purdom, *The Garden City*, p. 305.
[83] Ibid., p. 104.
[84] Ibid., p. 105.

in order, the cost charged to the lessee. Those who disliked gardening would have had to employ a gardener. However, according to Purdom a garden "is irresistible to the man of wholesome mind,"[85] and provided both individual and civic pride, as the mark of a "complete citizen."[86]

Gardening was understood as a vital leisure time activity, a relief from a hard day's work, combining the best of the country and of the city. At Letchworth gardening was such an important civic duty that there were formal classes in gardening offered to children at the school. The town was understood by Garden City advocates as a "community of gardeners."[87] A community brought together by gardening,

> and taught the virtues of patience and resourcefulness which come from the cultivation of nature, will, in the development of its social consciousness, acquire the strong qualities of mind and body which will fit it to experiment and adventure, without which our common life becomes stagnant. The occupations of the garden provide excellent training for the world and the government of affairs. They add to dignity and self-confidence, and cause men to think for themselves.[88]

In other words, the lessons learned in the solitary act of gardening were transferable to entire community, as a set of shared values; it is this balance between individualism and community that Howard argued for. Ironically, Purdom stresses that a garden is a private realm for cultivation and needs to be fenced or walled in from the public. However, the front garden cannot be treated as such and is a gesture to the community contributing to an overall sense that the city is in a garden. Here we see a clear distinction made between the role and functions of front and rear gardens. The front garden is typically the male domain with its emphasis on manicured appearance, while the rear garden is more likely to be the female realm, a wilder garden often devoted to growing vegetables and flowers. At Letchworth gardening was such an important civic duty that there were formal classes in gardening offered to children at the school. Purdom continues by stating that the "owner of this garden values everything in it and loves everything in it and would not exchange his quiet life there for all the excitement and profits of an uneasy world."[89]

By 1913, H.D. Pearsall, Chairman of the Howard Cottage Society Ltd., reporting on the building of cottages at Letchworth affirmed that gardening was actively pursued by residents of the working-class cottages. He wrote:

> All the cottages have gardens. Some are larger than others, to provide both for the enthusiastic gardeners (who are many) and for the few who dislike

---

[85] Ibid., p. 105.
[86] Ibid., p. 105.
[87] Ibid., p. 111.
[88] Ibid., p. 111.
[89] Ibid., p. 110.

gardening. But the average size is one-twelfth of an acre, and in the large majority of cases they are well cultivated and filled with flowers and vegetables, often yielding all the vegetables for the need of the family.[90]

Beyond providing a place for honest labor, gardens were also intended to provide working-class families with a significant amount of their annual food.

The controlled size of Letchworth, and the widespread cultivation of greenery, in both public and private spaces and at a variety of scales, led to the development of an environment characterized by what Purdom has described as "open space," a spatial system that is a refuge from the "restrictions of the town."[91] Purdom argued that this was a comprehensive spatial idea that pervaded the town, supported by the extensive role of green space, throughout the community. At Letchworth, green space became generalized.[92] The continuity of green space from outlying farm through to private gardens in town was proposed as a unifying factor, breaking down historic divisions and creating interconnectivity, and yet also providing legibility between elements. The radical notion that the city could become a community of gardeners, extended the idea of the rural village into a general model of urbanization. The preponderance of greenery and green space, provided an ecological overlay to the engineered infrastructural systems on and below the surface of the earth.

Beyond the private residential garden there are series of green space types that contribute the openness of the Letchworth plan, and the continuities of greenery. The center of the community is occupied by a small-town square (which has never been fully realized), formal green spaces along the Broadway, and conventional "high" streets with shops. The center is surrounded by various neighborhoods, areas for factory sites, and green spaces such as Norton Common. The residential neighborhoods (originally conceived by Howard as wards) support small public green spaces and the private gardens, finally the entire town is encircled by an agricultural belt (or greenbelt). All of this was intended to give the inhabitant the sense of living in the country, and in an environment that is healthy and clean. Therefore, five types of green space can be identified: 1) the agricultural belt which allows for the preservation of agricultural activities adjacent to the town and acts as a thick green boundary, and includes some institutional and recreational uses; 2) the central square and adjacent green spaces, which is not significantly different from a traditional town center; 3) public parks, such as Howard Park and Norton Common; 4) neighborhood common areas; and 5) private gardens modeled after the rural cottage garden. The major green space innovations were the agricultural belt, and the emphasis on private gardens and the act of gardening. Letchworth was conceived around the experience of being in a garden, at both the intimate level of the private garden and at the communal level of the town.

The various kinds of green space found at Letchworth tended to align with different kinds of gardeners, although the residential gardener remained the most

---

[90] Ibid., p. 262.
[91] Ibid., p. 112.
[92] Ibid., pp. 112-13.

prominent. Historically, there have been various types of gardeners involved in cities: 1) the gardener as a designer, someone who conceives of a garden (who has evolved into the landscape architect) or urban landscape; 2) the gardener as an administrator overseeing a park or green space system (the head gardener or parks planner); 3) the menial gardener in the employ of someone else, typically tending a garden or park for an aristocrat or a municipality; and 4) the private or individual gardener who looks after their own small private garden or allotment, typically adjacent to a house, or who maintains a communal garden such as found in a monastery.

In Letchworth we find all of these, the homeowner who gardened during "leisure" time, those who managed the municipal green spaces, and the designers of the community. This emphasis placed on the gardener by the early Garden City movement meant that the gardener emerged as an important urban figure, what was intended to be a counter to the more obvious figures of the *boulevardier* and the *flâneur* associated with the radical transformation of Paris during the latter half of nineteenth century, or the subsequent emergence of figures such as the *automobiliste* and athlete associated with various modernist visions of the city put forward during and after World War I, such as those developed by the Futurists and Le Corbusier. These nineteenth century figures represented the sophisticated urbanite occupying the traditional spaces of the city, such as the street and the square, while the early twentieth century figures represented the romanticism associated with the advent of mechanical technologies and the development of high volume transportation systems. Beyond the international Garden City movement, the gardener (or cultivator) would be vital to Le Corbusier's Ville Contemporaine project of 1922, which, although it focused transportation and athleticism, also included a Garden City element and provisions for gardening; Le Corbusier, while critical of the traditionalism of the Garden City movement, clearly borrowed ideas from it.[93]

Another, although unrelated, example, was Frank Lloyd Wright's agrarian city vision captured in his Broadacres project of the 1930s where homeowners were expected to actively cultivate one-acre parcels of land. More direct descendants of the Garden City such as the British New Towns developments after 1946, borrowed ideas but de-emphasized the role of gardening. The New Towns were in many ways closer to the post World War II suburbia, bedroom communities that lacked the integration and intended self-sufficiency of Garden Cities. The gardener, working in the relative isolation of a private plot, lacks the public presence associated with some of the urban figures noted here. The gardener remains in the background, in effect the missing figure in the development of the city in the twentieth century, which has derived so many ideas from the Garden City movement. Ultimately the gardener made a brief appearance and has been largely forgotten.

The garden was understood, by the advocates of the Garden City, as a complex space which integrated labor, country, town, pleasure, and moral and civil education. Gardening is a contemplative activity involving the creation of artificial ecologies, or assemblages, of plantings, spaces, and temporalities. As noted above, gardens are

---

[93] See M. Swenarton, "Rationality and Rationalism: The Theory and Practice of Site Planning in Modern Architecture 1905-1930," *AA Files* 4 (July 1983), pp. 49-59.

reliant on labor and produce many things. The garden assemblages at Letchworth operated closely with the house, to create a unit that varied in size, function, and economics, but were fundamentally the same, driven by the influential domestic design principles of the Arts and Crafts movement. These embraced a full vision of living that encompassed work and leisure, defining everything from furnishings, interior arrangements, morality, to the organization of plantings in the garden; these units aggregated into neighborhoods, and the town as a whole. The residential gardens at Letchworth, tended to be strongly bounded, and were subject to a gendered and functional difference between front and rear gardens. Despite some of the social theories embraced by proponents of the early Garden City, the organization of both interior domestic and garden space followed traditional gender lines with the kitchen and rear garden clearly defined as a women's realm, while men tended to the front garden.

As Kropotkin and Carpenter suggest the labor involved in gardening is a manual, intellectual, and creative activity; the creativity inherent to gardening is consistent with Deleuze and Guattari's emphasis on innovation.[94] To consider a Garden City, such as Letchworth, as a multitude of assemblages, refers to the machinic, linguistic, and territorial processes of the town, and the creative or productive aspect. The expressive aspect of the garden as an assemblage involves semiotic and linguistic factors, and in the case of Letchworth reflecting the collective aspirations of the community. These factors, along with the territorial innovations of the Garden City, had an innovative dimension that would inspire further experiments around the world.[95] As the Garden City movement stressed, the lower densities and economics of land development were shaped to give all houses a garden.

Gardens are usually defined by fences, walls, or hedges, however, there is always the potential to interconnect gardens into larger ecological systems. To some extent this was achieved in that successions of gardens can link together to create an ecological matrix, which also participates with the various public green spaces in the town and the encircling agricultural belt. The agricultural belt has created an urban growth boundary but did not fundamentally change the relationship of the town to the country as it was intended to do. Although many boundary systems were employed at Letchworth, these would have enhanced, impeded, and blocked a wide range of flows, including water, air, energy, nutrients, and organisms. A city or a town disturbs and modifies surface flows the most, having lesser effects on subsurface and supersurface flows.

While the Garden City was premised on a rich understanding of the garden and gardening, subsequent models both affirm and deviate from this agenda. With the radical broadening of green space systems, during the twentieth century, we can suggest that the gardener has had tremendous impact on the development of the city. And yet, gardening is largely a solitary and anonymous activity, the gardener remains an underappreciated figure despite its long association with the history of cities. The Garden City concept that the city could become a community of gardeners, extended

---

[94] See P. Hallward, *Out of this World: Deleuze and the Philosophy of Creation* (London: Verso, 2006).
[95] See, for example, Hall and Ward, *Sociable Cities*.

the idea of the rural village into a general model of urbanization, but it would also degenerate into a suburban condition where the garden is effectively no longer an essential element. To some extent this began with the design of the highly influential Hampstead Garden Suburb in London, founded in 1907, by Parker and Unwin, where the central role of gardening was largely abandoned. Further, the development of Garden Suburbs would undermine the self-sufficiency inherent in the Garden City model. Rather than cultivated landscapes, urban green spaces have become maintained landscapes. The notion of a community of gardeners has not been systematically developed as a concept. However, while the Garden City has often been linked to suburbia, as Lewis Mumford, and others, have argued the contemporary suburban model has fundamental differences from the ideals of the Garden City.[96] The Garden City was always conceived of as a self-contained settlement with all the amenities, not merely an appendage to a central city.

The city as a garden, comprised of gardens, remains a powerful analogue for the ecological, or sustainable, city. The gardener could reappear as a real urban figure, not an anonymous member of the maintenance crew. The extensive provision of green space, both at the public and private level, in the Garden City model meant that gardening was given a general role in society not seen before in an urban, or quasi-urban, model. Ebenezer Howard's vision of the Garden City has produced a complex legacy. On the one hand are the many green space typologies that have emerged since the nineteenth century including greenbelts, greenwebs, and greenways. On the other hand, the notion of a community of gardeners has been largely abandoned. If we consider the concepts presented above regarding agency against the operations of the early Garden City, we can make the following observations. The private garden is a complex and productive cultural and horticultural space, it is situated between the farm and the square and resonates with creativity, freedom, innovation, and cultivation. Conversely, urban gardens and parks blend individual, bureaucratic, and self-ordering types of agency. Since the nineteenth century cities globally have increasingly incorporated a range of green space typologies from the private garden, comprehensive park systems, to encircling greenbelts. Each typology captures some aspect of the primal garden, these spaces have become part of the complex ecologies of cities and will increasingly need to be integrated as cities strive to find new ways of addressing the current environmental crisis.[97]

GREENBELTS AND THE ROLE OF THE FARMER

The role of farming in the Garden City model was intended to be a vital one, a key part of the attempt to unite the country and the town. With similarities to gardening, farming focuses on the production of food as it typically occurs at a larger scale and lacks, since the emergence of industrialization, the complex cultural associations

---

[96] See L. Mumford, "The Garden City Idea and Modern Planning," in E. Howard, *Garden Cities of Tomorrow* (London: Faber and Faber Ltd., 1945).
[97] See, for example, Galen Granz, *The Politics of Park Design: A History of Urban Parks in America* (Cambridge, Mass.: MIT Press, 1982).

linked to gardens. Nevertheless, farming requires specific methods, tools, and infrastructure, and has its own history. From 1880 until the outbreak of World War I, agriculture in Britain experienced a depression, as cheap produce was available from abroad, this resulted in fewer farms, a reduction in agricultural land, and the implementation of new practices.[98] During this period of agricultural uncertainty the Garden City movement emerged. At Letchworth a concerted effort was made to involve (and reinvigorate) the agricultural community that was both displaced by the town and surrounded the town. This was attempted at Letchworth by establishing a community that, through various mechanisms, including providing a local market for farm produce, was integrated with agriculture; however, this met with little success.[99]

Letchworth, while taking advantage of cheap agricultural land and depressed farms, was intended to provide a stable market for local agricultural producers, both those occupying the agricultural belt and paying rent to the Garden City company, and those farmers in surrounding areas. The agricultural belt encompassed land for farms and allotments under the ownership of the company, providing an inter-zone between town and country. While widely adopted during the twentieth century, the agricultural belt, or greenbelt, has not emerged as a strategy as agriculturally productive as originally envisaged. Beyond preserving agricultural and undeveloped land and providing a green edge to cities it remains a rather static urban device.

Between 1700 and 1880 there were many significant developments in British agriculture, including the systematic enclosure of land, the advent of mechanized farming, developments in land reclamation, new crops, and new practices in land and livestock management that produced greater and more reliable yields. The advances in farming techniques during this period were consistent with the general technological advances of the Industrial Revolution. Before the enclosure of land in Britain beginning in 1760, agriculture was more flexible and open to those without significant capital, as Orwin notes:

> The flexibility of farming in the open fields produced holdings of all sizes [and shapes] that could be broken up and reassembled to suit the changing capacities of the various members of the community.[100]

As he points out, this pre-industrial system resulted in farmers highly knowledgeable in the manual aspects of farming, but less knowledgeable about the science and economics of agriculture.[101] The application of both industrial and scientific methods to farming radically transformed agricultural practices during the nineteenth century. Further, forces arose during this period that transformed agriculture into a global system. The many continuing challenges faced by farmers tended to make them cautious as a group, averse to risk. The plight of rural workers, many of whom were driven into the cities seeking employment, was a central aspect of the Industrial

---

[98] See C.S. Orwin, *A History of English Farming* (London: Thomas Nelson and Sons Ltd., 1949).
[99] See M. Miller, *Letchworth: The First Garden City* (Chichester: Phillimore & Co. Ltd., 1989), pp. 136-141.
[100] Orwin, *A History of English Farming*, p. 110.
[101] Ibid., p. 111.

Revolution, ameliorating the resulting living conditions in early industrial cities was a major preoccupation of the Garden City movement.

Among the major aims of the Garden City the revitalization of local agriculture and improving the standards of living for farm workers (together with those of industrial workers). Orwin divides these into "servants" who typically lived on the farm and enjoyed a degree of employment security, and day laborers who were employed as required. This applied to both men and women and resulted in different types of work between the two groups. The male servants were "the carters, waggoners or horseman, the shepherds and the cowmen" while male and female day laborers were involved "with field work of all kinds, hedging and ditching, haymaking and harvest dung-spreading, and so on."[102] Female farm servants "were occupied mainly with work in the house or dairy, but milking in the fields and cowsheds, and tending calves, pigs and poultry, came also within their sphere."[103] As Orwin points out farm workers tended to be behind industrial workers with respect to employment standards, particularly day laborers.[104] This led to the formation of the National Agricultural Labourers' Union in 1872, however, it struggled to be an effective agent of reform. By the twentieth century this tradition had largely disappeared in Britain, replaced by the wage-earning farm worker.

In the Introduction to his text, Howard acknowledged the challenges of restoring people to the land against the strong attraction of the city. Within the limits of the town various kinds of agricultural enterprise were expected to occur within the agricultural belt, including large farms, small-holdings, and allotments. He wrote:

> the natural competition of these various methods of agriculture, tested by the willingness of occupiers to offer the highest rent to the municipality, tending to bring the about the best system of husbandry, or, what is probable, the best systems adapted for various purposes. Thus, it is easily conceivable that it may prove advantageous to grow wheat in very large fields, involving united action under a capitalist farmer, or by a body of co-operators; while the cultivation of vegetables, fruits, and flowers, which requires closer and more personal care, and more of the artistic and inventive faculty, may be best dealt with by individuals.[105]

As Howard noted, a primary purpose of the agricultural belt was to produce revenue for the town through the charging of rents. This would be higher than before the town was established due to the creation of a local market, and the provision of fertilizer in the form of treated sewage from the town; this would create a cycle of wealth that involved improving land and a captive market for produce. Howard believed that farm tenants in the agricultural belt would be prepared to pay a higher rent due to the situation.

---

[102] Ibid., pp. 94-95.
[103] Ibid., p. 95.
[104] Ibid., p. 97.
[105] Howard, *To-morrow*, p. 17.

In 1903, the proponents of the first Garden City were able to assemble a group of adjoining parcels of land (estates) at a good price for the subsequent development of the town. The land acquisition was carried out in a discrete and quick manner. On October 9, 1903 a formal opening was held which included shareholders and various dignitaries. The estates that were purchased were occupied by various tenants, so that as the development of the town progressed these agreements were managed in order to minimize revenue losses.[106] During the first year's relations had to be established between the company (that had no public authority) and surrounding towns and exiting local authorities. For example, the land that was purchased was part of three existing parishes (Letchworth, Norton, and Willian); in 1907 these were amalgamated into a single civic parish. Despite the best of intentions, the introduction of a new town into a quiet and undeveloped rural area, created tensions between the Garden City and local communities.[107]

On September 10, 1904 a conference was held at Letchworth on the subject of the "Garden City in Relation to Agriculture," at which Thomas Adams delivered a paper, and a number of key people were attendance, including Ebenezer Howard. Mr. Alderman Winifrey made remarks regarding experiments in putting people back onto the land in Lincolnshire, beginning in 1899. Those experiments involved farm laborers becoming tenants of small holdings, much as proposed at Letchworth, which had proved after five years to be very successful in retaining people on the land.[108]

The plan for Letchworth called for a variety of scales of agricultural operation; studies of practices in Denmark, by the Garden City group, affirmed the concept of small holdings, which were seen as a way for some citizens to be self-supporting.[109] In his book *Garden City and Agriculture: How to Solve the Problem of Rural Depopulation* published in 1905, Adams outlined the issues surrounding rural depopulation, and defines the basic elements of the Garden City; he discussed how the Garden City would revive the local agriculture "by bringing a market to the door of the farmers, providing security of tenure, establishing small holdings, promoting co-operation, and giving the laborer accessibility to the social attractions of the town."[110] Adams argued that urban over-crowding is more a problem of poor planning, than of urbanization per se, he wrote: "The fact that people huddle together in slums is not a result of industrial concentration; it is due to the fact that the concentration has not been properly controlled or directed."[111] Further, Adams stated that the two primary causes of over-crowding in cities are "the lack of public control over the building of towns, the planning of streets, the provision of air space, etc." and the "scarcity and dearness of land, due to the system of land tenure and the natural selfishness of the urban landlord."[112]

---

[106] Purdom, *The Garden City*, p. 46.
[107] Purdom, *The Building of Satellite Towns*, p. 144.
[108] See Appendix I in Thomas Adams, *Garden City and Agriculture: How to Solve the Problem of Rural Depopulation* (Hitchin: Garden City Press Ltd., 1905), pp. 115-117.
[109] See Appendix IV in Adams, *Garden City and Agriculture*, pp. 137-144.
[110] Ibid., p. 14.
[111] Ibid., p. 19.
[112] Ibid., p. 20.

Laborers leave the country for the town, because they want improvement for themselves and their children, the laborer "is educated above the petty tyranny of the village, the long unbroken hours of labour, the lack of opportunity, and the insanitary cottage."[113] At the beginning of the twentieth century in Britain there were many factors that contributed to the agricultural depression, including the lack of capital, the lack of education in farming methods, the lack of good affordable housing, and the loss of traditional small rural industries.

The plan for Letchworth does not go as far as integrating farming into the structure of the town, although the growing of food was intended to occur in private gardens and in the agricultural belt. However, the presence of farming was evident in the compact size of the town, and the proximity of the surrounding farms to all parts of the town. This was particularly the case in the first decade as the town developed in a relatively haphazard way. By 1913 the agricultural belt, which originally comprised two thirds of the entire land area, supported 75 tenants working parcels of land of varying sizes; the rates of return and condition of the agriculture were deemed "satisfactory."[114] As Purdom stated the Garden City "does not, like other towns, destroy rural pursuits; it intensifies them."[115]

Beyond the production from residential gardens and allotments, Purdom described the produce from the agricultural belt in 1949 in the following terms:

> The chief products of the agricultural belt are fruit and dairy produce. A small amount of land is used for market gardening; but the land is not specifically suitable for this purpose, being, in the main, rather exposed and needing considerable 'making.' Apples, pears, and soft fruit are grown in quantity for the local market. It is not possible to tell what precise extent the demands of the new town are met by the produce of the agricultural belt. That the area does not meet the needs of the town is certain.[116]

Despite this, Purdom also suggested that the agricultural belt was an essential part of Letchworth's economy and structure, and that by 1949 the original goals for the belt had been achieved, he wrote:

> The effect of the garden city on agricultural life can already be seen at Letchworth, although the rural belt is the least developed part of the town. The agricultural population has been increased, the cultivation of the soil has enormously improved, wages are higher, and the condition of the farm worker, his outlook and interests, have been raised to the level of those of the town worker. On the other hand, the townsman and factory worker have been brought into immediate touch with the country.[117]

---

[113] Ibid., p. 26.
[114] See H. Burr, "Agriculture and Small-Holdings in Garden City," in Purdom, *The Garden City*, pp. 272-283.
[115] Purdom, *The Garden City*, p. 116
[116] Purdom, *The Building of Satellite Towns*, p. 145.
[117] Ibid., p. 146.

Conclusion: Letchworth after 1913

At Letchworth the designers and builders of the town attempted to produce a synthetic model that addressed a broad range of urban issues from structure, spatial organization, housing design, civic education, to political, social, and cultural factors. As an experiment in the design of a total community it was relatively successful, despite the many years it took to reach its target population. The agricultural belt at Letchworth has been an enduring and important aspect of the town since its inception. However, the agricultural and economic impact of the belt has waned over time. Writing in 1945 F.J. Osborn stated:

> For Letchworth was, and remains, a faithful fulfillment of Howard's essential ideas. It has to-day a wide range of prosperous industries, it is a town of homes and gardens with ample open spaces and a spirited community life, virtually all its people find employment locally, it is girded by an inviolate agricultural belt, and the principles of single ownership, limited profit, and the earmarking of surplus revenue for the benefit of the town have been fully maintained.[118]

While not originally defined by ecological, or sustainable, design principles, there is no doubt that the Garden City model codified many aspects of twentieth century urban planning and design, and many of the innovations associated with the concept are now considered to be essential in contemporary sustainable urban design.

Purdom argued that this was a comprehensive spatial idea that pervaded the town, supported by the extensive role of green space, throughout the community. The continuity of green space from outlying farm through to private gardens in town was proposed as a unifying factor, breaking down historic divisions and creating interconnectivity, and yet also providing legibility between elements. The preponderance of greenery and green space, provided an ecological overlay to the engineered infrastructural systems on and below the surface of the earth. The advent of many new green space typologies during the twentieth century, inspired by the concept of the agricultural belt, has done much to reorganize the relationship of a city or town to its region, and also would aid in the significant impact of green space networks on urban structures and organizations.[119]

As Purdom stated, Letchworth was conceived around the experience of being in a garden,[120] at both the intimate level of the residential garden and at the communal level of the town. Further, the use of the agricultural belt, or greenbelt, as a wide boundary separating the town from the country and providing a protected zone for agricultural and cultural amenities, helped create the sense of a town engulfed in greenery. The stated aim of the Garden City movement was to unite city and country,[121] to draw from the best of the two worlds, and to break down what was seen

---

[118] F.J. Osborn, "Preface," in Howard, *Garden Cities of To-Morrow*, p. 13.
[119] See Cranz, *The Politics of Park Design*.
[120] Purdom, *The Garden City*, p. 113.
[121] See "The Three Magnets" diagram published in Howard, *Garden Cities of To-morrow*, p. 46.

as an artificial divide; in other words, to activate or remove the historic boundary between town and country, allowing for a greater range of interflows between the two. The use of the agricultural belt as a wide boundary connecting the town to the country and providing a protected zone for agricultural and cultural amenities, helped create the sense of a town engulfed in greenery. This is augmented by other green spaces in the town itself and enhanced by the generally single-family housing fabric and the emphasis placed on private gardens. However, despite the successes of various greenbelt initiatives in Britain, Canada, and the United States the role of the greenbelt has been largely reduced to an urban growth boundary and as a mechanism for preserving landscapes. The economic import placed by Howard on the original concept has been effectively rejected; this is evident in the transformation of the original agricultural belt to the greenbelt. Nevertheless, the host of green space typologies that have emerged as a result of the Garden City provide a whole range of related urban amenities. The influence of Howard's theories has been widespread, and continues to be felt, despite the fact the Garden City as a total concept, agricultural belt included, has been largely abandoned.

The early Garden City was in many ways innovative; as noted above the legacy has had both positive and negative effects on subsequent urban development. The efforts of the founders of the movement drew together many influences, as Howard so transparently demonstrated, resulting in many productive concepts. Letchworth was based on a set of ideas and diagrams drawn together by Howard and implemented by others. As the historical record demonstrates the first decade involved much trial and error on the part of the founders. Howard's model synthesized many factors into various diagrams that produced many outcomes, some short-lived and some longer term. Fundamentally, the Garden City attempted to develop a new town, a community that reconceived the inter-connection between town and country and many of the commonly held assumptions about urban design. There is no doubt that Howard proposed comprehensive social reform, however, as the Garden City movement developed, the focus narrowed to that of physical planning, which remains a problem with urban planning and design today. Further, the Garden City, even as manifested at Letchworth, was a total concept. However, much of the enduring legacy of the movement has been the ideas selected from the model, often inappropriately applied.[122] This has included aspects such as the Garden suburb, the greenbelt as a device for bounding cities, and the emphasis on the single-family house and garden. Each of these concepts, when poorly applied, has often resulted in exacerbating urban problems.

A number of proponents and historians of the Garden City movement have attempted to define the innovative aspects of the model. As Osborn stated, these include the commitment to relocating populations from both the city and the country to the Garden City, limiting the size of the community, providing a quality of living (including housing, open space, recreational and cultural opportunities) not found in either the city or country for those of modest means, addressing the historic rift

---

[122] See S.V. Ward, "Ebenezer Howard: His Life and Times," in Parsons and Schulyer, eds., *From Garden City to Green City*, pp. 32-33.

between urban and rural areas, providing comprehensive mechanisms for the planning and management of towns and regions, conceiving of a community that functions form the micro (house and garden) to the macro (region) scale in a continuum, providing a common ownership of land that benefits all, and creating a local and shared form of governance. [123] From Osborn we can suggest that the comprehensive planning of a town remains an enduring legacy of the Garden City, an almost obsessive control of all aspects of the design from interior space design, architectural controls, careful infrastructure design, the design of vistas, to the functional organization of the town. These would become the hallmarks of twentieth century, often resulting in over-planned environments. It is worth noting that Osborn fails to mention the importance of green space in the design of Garden City.

Writing in 1949, C.B. Purdom discussed the successes and failures in the development of Letchworth. The location of the town presented challenges, and the financial structure of the development was always tenuous as Letchworth was a private venture. Purdom suggested that the board of directors was often fragmented and lacked a concerted direction, resulting in inertia and poor administrative mechanisms.[124] Despite a lack of presence in the public imagination after World War I enough was realized at Letchworth in the first decade to demonstrate the fundamental concepts of the Garden City. Ward points to what he considers a series of failures in Howard's vision, these include: the inability to establish a system of collective ownership of land, the lowering of density at Letchworth from Howard's original proposal, the emphasis at Letchworth in planning over community building, the social hierarchies employed in the organization of Letchworth, and the functional zoning of the town. [125] There is no doubt that while the original Garden City movement realized only two towns, Ward also notes that the Garden City model had strong international influence on planning from the 1920s to the 1950s, when it came under attack, from figures such as Jane Jacobs. More recently it has come back into favor with the rise of New Urbanism on one hand, and the move to sustainable communities on the other.

After the initial decade the development of the town was interrupted by World War I, and difficulties in securing financing remained. Eventually the establishment of a second town at Welwyn Garden City in 1920 would also divert attention from Letchworth. Slowly, Letchworth became more conventional in its operation, although this is also due to the fact that many ideas developed at Letchworth became common practice. Mervyn Miller addresses the agencies required for the establishment of the community, including the early investors, designers, and pioneering settlers, and also underscores the notion that the pragmatic tone of Howard's writings was an essential element in the enterprise; in other words, as an assemblage his book was highly successful.[126]

---

[123] Osborn, *Green-Belt Cities*, pp. 32-33.
[124] Purdom, *The Building of Satellite Towns*, p. 174.
[125] S.V. Ward, "The Howard Legacy," in Parsons and Schulyer, eds., *From Garden City to Green City*, pp. 222-229.
[126] Miller, *Letchworth*, pp. 210-211.

Walter Creese's assessment includes many of the visual factors that influenced the design of early Letchworth, but he also focuses on functional and structural aspects of the design.[127] The emphasis on careful design of the site, the planning of infrastructure, the consideration of the relationship between house (individual) and town (collective), and between town and country, were all vital innovations. Creese underscores the aesthetic and site-specific aspects of Letchworth's design, the detailed attention paid by the designers to picturesque compositional affects, or the high degree of integration across scales. Discussing the choice of architectural style by the various architects involved in early Letchworth, Creese writes that the town "was architecturally an event in the hunt for a more deeply rooted and authentic folk life."[128] Parker and Unwin's house designs drew from the English cottage tradition, and focused on simple execution, strong connection to environment, and careful design of the living environment. The Parker and Unwin firm were very committed to the design of quality and healthy housing for the poor. This included a strong relationship to the outside. For example, Unwin discussed the value of gardening in an 1897 speech, when he stated that gardeners through the act of gardening often receive "more education than from many books."[129] The housing of the working classes in cottages (attached and detached), rather than in higher density forms of housing was integral to the entire design of Letchworth. The houses had to be inexpensive, functionally planned, minimum in area and volume, and undecorated. In their essay "The Art of Building a Home," written by Parker and Unwin in 1901, they addressed functional planning, art and simplicity, practical details, and the harmonious clustering of groups of cottages.[130] Both Unwin and M.H. Baillie Scott advocated for a single, simple, and flexible living space in the design of small cottages to accommodate family functions, with separated bedrooms. This is further testimony to the notion that Howard's text was a starting point for an exercise that allowed various contributors to innovate, particularly the Parker and Unwin practice. There is also the question as to whether or not the Arts and Crafts vision employed at Letchworth, with its nostalgic overtones and picturesque qualities, was really consistent with Howard's vision?

Among the many innovations developed at Letchworth, are those such as agricultural belt (green belt) that have been widely discussed, and those that have not been as extensively examined, such as the role of gardening in the development of the model. The emphasis on a range of green space types was implied by Howard, but fully developed in the Parker and Unwin scheme. This meant that the gardener, always a presence in cities, was to take a central role. Some recent authors note that the Garden City is a precursor to contemporary interests in the sustainable, or ecological, city.[131] Features such as the agricultural belt (growth boundary), compact size, good transit linkages, abundant green space and vegetation, agriculture, gardening, neighborhood design, and site-specific design are all consistent with

---

[127] Creese, *The Search for Environment*, pp. 169-173.
[128] Ibid., p. 214.
[129] Quoted in Meacham, *Regaining Paradise*, p. 85.
[130] See Creese, ed., *The Legacy of Raymond Unwin*, pp. 47-54.
[131] For example, see Hall and Ward, *Sociable Cities*.

current thinking. However, other factors, such as the low density of the overall development, have consistently linked the Garden City to the subsequent worldwide development in suburban environments. As many advocates have argued, the Garden City was intended to be a self-sufficient satellite community, not merely a bedroom community. At Letchworth this is clearly the case, however, in subsequent developments, such as Hampstead Garden Suburb and Welwyn Garden City, this no longer was upheld. However, like any urban model, and particularly one that has had such widespread influence on urban developments throughout the twentieth century, it is important to test some of the original assumptions against contemporary ideas of innovative productivity (assemblage theory) and urban ecology.

The structure of Letchworth as a planned town involved the careful organization of land by function, the employment of a wide range of boundary systems, and the management of flows through infrastructure. All of this was integrated and used advanced approaches for the day, and in a number of areas was pioneering. The idea of the "patch" as a defined piece of land, one that is bounded, and has a uniform structure is similar to concepts employed in modern urban zoning which define parcels of land by function, often resulting in urban environments that tend to be a patchwork of functionality, however, the overall patchiness of a system, and its latent structure as a patchwork, depends upon composition and scale. Cities define territories by infrastructure (roads, canals, etc.), property, and topography in a manner that tends to be rigid, as opposed to the open and fluid territorial systems of nomadic cultures. Nevertheless, there has always been some kind of functional organization in cities, which separated religious, administrative, residential, and industrial functions, however, the systematic zoning of cities is a more recent phenomenon.

CHAPTER 10

Conclusion

The original plan for Letchworth Garden City by Barry Parker and Raymond Unwin, was selected, in part, because of its careful attention to the local site features, including topography, drainage, solar orientation, existing features (railway line, villages, trees, etc.), and soil conditions. These were taken into account with respect to the functional organization of the town, orientation, and the location of services and infrastructure, including water supply, sewers, and waste disposal. The commitment on the part of the early Garden City movement to the widespread use gardens, parks, and plantings, meant the whole town was indirectly conceived of as an artificial ecology. For example, the planting of a wide variety of street trees was a feature of the town, however, as these matured the tree canopy became a source of complaint for residents whose sunlight was often blocked out. The design of Letchworth also accommodated the orientation of buildings (and rooms) and gardens to optimal sun exposure, the conservation of topsoil, the protection of existing trees, and the care of gardens and green spaces.[1]

As Letchworth was developed as a new community it could be conceived of in its entirety. Utilizing infrastructure systems available at the time, and incorporating a full range of green space systems, it can be argued that it was an attempt to conceive of a town as a fully functioning ecology. The attempt to put in place a community based on a unified ownership allowed for a shared form of governance. Letchworth was conceived of as a garden, the plan by Parker and Unwin created a continuum from the space of the private garden to the scale of the region, one that included the neighborhood, the town, and surrounding farmland. The use of zoning as a method for functionally organizing the town, and the integration of modern infrastructure, was often circumvented by the green space systems, and flows such as wind and water, which as Deleuze and Guattari suggest cannot be striated.

The emphasis on the single-family house, the private garden, and the civic act of gardening, were key concepts for the Garden City. Here, the historic role of the gardener is reterritorialized in the Garden City.[2] The garden, which is located midway between the farm and the urban square or street, unites agriculture (the farmer) and culture (the urbanite), through cultivation. It is a highly productive, and complex space. The gardener is attuned to the seasons, and to some extent lives a cyclical life, however, gardeners also know that every year is a new set of circumstances with new opportunities and challenges. The garden is the space in which urban living coalesces, it is the shifting space of love, botany, discourse, philosophy, art, and entertainment. It

---

[1] F.J. Osborn, *Green-Belt Cities: The British Contribution* (London: Faber and Faber Ltd., 1946), p. 75.
[2] See A. Parr, "Deterritorialisation/Reterritorialisation," in A. Parr, ed., *The Deleuze Dictionary* (New York: Columbia University Press, 2005), p. 69.

is a territory that is constantly deterritorialized and reterritorialized by the seasons and the weather, by the labors of the gardener, and through the pleasures it affords to those that use it. Thunder storms, rain showers, frosts, and other forces of climate each deterritorialize and reterritorialize the space of the garden, as do the plants emerging, blooming, and retreating during their annual cycles. The space of a garden, while descended from a tendency to striate and bound space, is productive in many ways. The gardener, in an ever-changing and creative effort, to reconceive the garden, engages in a living space. Gardening creates fertility of the soil, often enhancing the ecological diversity of a site.

The notion, put forward by the early Garden City movement, that the city could become both a garden and a community of gardeners, continues to be a model for thinking about the creation and maintenance of ecologies inhabited by humans. While many aspects of the original Garden City have been widely adopted, there are also notable failures associated with the model. The original model was driven by social and structural ideas for integrating agriculture (the country) and the city and was not primarily focused on ecological factors. The suburban qualities often attached to the Garden City continue to obscure other achievements: development of the greenbelt and numerous other green space typologies, and the focus on gardening as collective and spatial activities. Despite the fact that the early years of Letchworth attracted many committed settlers, the role of gardening would not become more vital than that found in any other small British towns. And while Letchworth today bears the hallmarks of a remarkable experiment, it struggles to define itself as a community.

The sustainability of the early Garden City scheme is reflected in various features of the design, including that it was a compact and self-sufficient community, it had relatively high densities (by contemporary suburban standards), it created a comprehensive open space system, and it was intended to be part of a cluster of small cities (the "Social City").[3] Hall and Ward have examined the history of the Garden City movement, the history of British planning, the New Towns after 1946, and recent developments in sustainable urban design in Britain. They identify twelve "key strategic policy elements" for sustainable cities, some of which resonate with the early Garden City, although they also recognize that there have been substantial changes in rural and urban issues since the late nineteenth century. The twelve elements are:

1. Develop Urban Nodes.

2. Selective Urban Densification.

3. No Town Cramming.

4. Strategic Provision for Greenfield Development.

5. Distance.

---

[3] P. Hall and C. Ward, *Sociable Cities: The Legacy of Ebenezer Howard* (Chicester: John Wiley & Sons Ltd., 1998), p. 23.

6. Top-Quality Linkages.

7. Clustered Development:

8. Town Expansions and New Towns.

9. Density Pyramids.

10. Variation According to Geography.

11. Areas of Tranquility.

12. Stimulate Remote Rural Areas.[4]

These twelve elements include: the notion of concentrating development around transit hubs; the protection of green space including agricultural areas; the siting of new towns a minimum distance of 50-85 miles from a major center such as London to ensure a degree of autonomy and self-sufficiency; the provision of well-connected and efficient transportation systems, preferably public; recent research confirms that small communities the clustering of smaller compact communities along public transit links is the most sustainable urban form as confirmed by research (Howard's Social Cities concept); building up density towards transit hubs; taking into account the specifics of a site and its topography, geology, orientation, etc.; and establishing a good working relationship between a city and the surrounding countryside. Further, the focus on interconnectedness, via rail and road, was crucial to the selection of the site for Letchworth, the bounded community with a transit node in the middle of it foreshadowed the current focus on transit-oriented development. The original agricultural belt was intended to give communities a distinct identity by preventing uncontrolled sprawl and the merging of towns and cities into each other. The concept of clustering groups of distinct communities into one super community, linked by rail, replicates the notion that cities are often collections of villages, and foreshadowed the post-World War II plan for a city like Canberra, Australia and its use of green systems. Letchworth was precisely designed to respond to its site, and to stimulate agricultural activity in the surrounding areas.

A key innovation of the Garden City movement is the concept of the agricultural belt, which transformed into a wide range of new urban space types. The agricultural belt was proposed by Howard as a spatial interface between the rural and urban aspects of the Garden City, to remove the divide that exists between urban and rural populations. Farmers often have been seen as unsophisticated, uneducated and unworldly by city dwellers. Farmers are strangers in the city, seemingly lost and out of place. And yet, as Howard implies, the two populations, rural and urban, form part of a single territorial and operational enterprise, and cannot exist without each other.

---

[4] Ibid., pp. 151-154.

This reflects a regionalist approach to the design, construction, and management of cities. The farm and farmer were the product of urbanization, and along with the city, create the concept of the region; innovations in agriculture have always come from the city, against the conventional thinking that cities are products of agricultural settlements. Ultimately, cities, agricultural lands, and comprehensive trading networks are part of one large complex system.

If we look at cities as assemblages of assemblages, or the interlocking and overlapping operations of numerous abstract machines, it is evident that any complex system, such as a city, will be to some extent self-ordering. In a city, as noted above, there are the multitude agencies and agents that could potentially contribute to the activation of the city as a complex set of ecologies operating on a wide range of scales and temporalities, but these tend to perpetuate an existing system. Landscape ecology, stresses both the structure of landscapes and the many forces, natural and artificial, that impact on how a landscape functions. As Richard T.T. Forman and Michel Godron argue, cities, as ecologies, suffer from a high degree of territorial fragmentation (typically from corridor or transportation systems), or patchiness. This results in a lack of ecological flow across urban landscapes outside of established channels, and this tends to make cities ecologically ineffective. Heterogeneity, diversity, flexibility, flow, and evolution within a landscape are key indicators of ecological functioning. Landscapes can transform quickly, as in the case of a natural disaster, or slowly, as in the case of a city. The stability of a landscape is dependent on many factors, including resistance and the ability to recover from changes; landscapes tend to oscillate between stable and unstable conditions, but ultimately move towards a "dynamic stability," seeking an ecological balance. In order to become more effective as eco-systems, or to reduce consumption and waste, cities need to be better harmonized with the ecologies they participate in; cities need to have more interconnected flows of resources, energy, organisms, information, nutrients, etc.

Traditionally, cities have been highly artificial environments reliant on largely mono-functional, channelized, and non-integrated infrastructure. Together with the heavy modification of landscape and ecologies caused by urbanization this has led to environments that are intensely interrupted and not ecologically effective. The contemporary city, which has embraced many aspects of the Garden City, therefore, becomes a territory for both, for the more rigid striated lines of the state and the city, and the nomadic lines of flight. The act of gardening was used by the movement as a vital concept for developing community and civic responsibility, and as a leisure-time activity designed to give pleasure and moral purpose. As the concepts of the movement spread, they transformed into the ubiquitous patterns of suburban sprawl, although devoid of many of the guiding ideas. Contemporary cities need to achieve a dynamic, and sustainable, balance between producers, consumers, and decomposers within complex eco-systems. The early Garden City provided many answers to the questions posed above, it created the model of a total community, where the whole city is a garden, connected to the larger gardens of agriculture.

Historically, urbanization and agriculture, organized land, distributing it among people in regulated ways through state systems, however, many flows such as those of information, capital, water, air, energy and subsurface movements cannot be so easily striated.[5] While cities encompass a multiplicity of assemblages, the ability to deterriorialize and reterritorialize themselves according to the potentials, intensities, and gradients in the structure, can be highly restricted. This requires boundary transformations or functional change, or the introduction of fluidic conditions. However, there are smooth space tendencies in contemporary cities that have potential for integrated urban/rural environments, including the emergent and everyday activities of systems. A territory, or assemblage, may or may not correspond with generally accepted (bounded) structures such as states, provinces, cities, or districts, as we know, many of these are arbitrary, and tend to be redefined by the agencies of a multitude of social, political, cultural, and governmental organizations.

While cities are highly striated, assemblages continuously operate, some of these creating smooth space conditions, others redefining historic urban structures and relationships. In many contemporary cities a wide range of nomadic forces have been at play including mechanical methods of transportation and electronic communications systems, that contribute to territorial instability. Human settlements, since the advent of cities and farms, have heavily modified the surface of the earth, producing territorial systems of varying degrees of intensity and functionality, usually bounded by infrastructure and political/economic structures; this includes field/road/irrigation ditch systems found in agricultural landscapes, and the higher density urban block/street systems in inner cities. In the suburban and peri-urban conditions the structure becomes more irregular and ambiguous. These developments in the last century have radically transformed the functioning, or performance, of landscapes, and have extended into the alteration of the subsurface and super-surface ecological flows and territorialities. The ragged edges of the contemporary city, and the emerging continuous gradient between agricultural hinterland and urban core means that urban ecologies can function in potentially integrated ways, allowing for fewer rigid boundaries within urban-rural landscapes. This regional entity, as Mumford writes, does not have "definite physical boundaries" and is a "system of inter-relationships that overflow and become shadowy at the margins."[6]

Whether urban ecologies are operationalized by the actions of individuals, bureaucracies, or the system itself depends on the diagrams of all the assemblages at play. Ultimately, the ecology of cities will depend upon all types of agency outlined above. The stability of a landscape, and the numerous boundary systems within a landscape, is dependent on many factors, including resistance and the ability to recover from change. Landscapes tend to oscillate between stable and unstable conditions, but ultimately move towards a dynamic stability, or seeking an ecological balance.[7] Cities, as complex ecologies, operate across a wide range of scales. Like ever-changing weather patterns, cities, are constantly shifting forces, many of which are self-directing, or emerge from a coalescing or collapsing of forces. The territorial machine provides a surface upon which the operations of the state, capital, and the

---

[5] Gilles Deleuze and Félix Guattari, *A Thousand Plateaus: Capitalism and Schizophrenia* (Minneapolis: University of Minnesota Press, 1987), p. 441.
[6] Lewis Mumford, *The Culture of Cities* (New York: Harcourt Brace Jovanovich, Inc., 1966), p. 315.
[7] Richard T.T. Forman and Michel Godron, *Landscape Ecology* (New York: John Wiley & Sons, 1986), pp. 431-435, 449.

like unfold. It is also the ground of nomadicism which operates in a fundamentally different way from the State. Territoriality in nomadic societies, or smooth space, functions more like an effective ecological system, whereas the operations of the state tend to stabilize and commodify land, to subject land to extensive geometrical subdivision, resulting in poor ecological flows.

The diagrams that define urban assemblages range from stable to unstable, from progressive to conservative, and these will determine the ability of a city to evolve its ecological performance. The arrangement of forces, organisms, territorialities, and languages that constitute an assemblage can be shaped or triggered by human agents. Ideally, these would operate in a nomadic mode, largely unconstrained by the bounded conditions of state systems, able to smoothly operate across the regulatory regimes of bureaucracies. For example, following the Garden City model, the gardener provides a potential model as an individual agent. At a larger scale, urban bureaucracies require dynamic models and the ability to radically modify urban boundaries. Depending upon cities to benefit from self-organization may occur in heavily blighted, or abandoned, cities but cannot be relied on in highly striated urban environments. It may be the case that systemic forces and the actions of individual can challenge or complement urban bureaucracies. Ecologically effective cities can be developed by activating the inherent patchwork structures that cities employ, in other words moving them towards smooth space through the modification, shifting, or removal of boundaries.

Whether a complex assemblage system, like a city, can effectively reorganize, or whether the intervention of agents and agencies is required, nevertheless, suggests that complex models must come into practice. These will only be effective if boundaries are modified, moved, or eliminated both within the city itself and within the agencies that manage cities. As John Rajchman notes the assemblage theories of Deleuze and Guattari are most effective when they make "visible problems for which there exists no program, no plan, no "collective agency," problems that therefore call for new groups, not yet defined, who must invent themselves in the process."[8] The agencing properties of assemblages depend on the ability of a given assemblage to be productive, in ecological terms this could mean that human interventions contribute, rather than detract, from environments.

The urbanite tends to live a striated condition or a system of relatively stable territories, these are continuously subjected to territorial forces that belong to individuals, collectives, agencies, bureaucracies, trades, professions, clubs, etc. Territories are also effected-affected by structures, intensities of incorporeal forces, weather patterns, geological forces, infestations, etc. The ability to transform an urban territory meets resistance due to the striated structure of land. However, there are special agents and/or intensities in the system that open up the potential for territorial transformation. The agents of deterritorialization may forge through a territory encountering significant resistance or a path of least resistance in the structure; a coalescing of determined or undetermined forces may create the same effective or affective potential. The systems that can perform the processes of deterritorialization

---

[8] J. Rajchman, *The Deleuze Connections* (Cambridge, Mass.: MIT Press, 2000), p. 8.

and reterritorialization include a whole host of forces that can come together to create an assemblage, these forces can be human actions (typically bureaucratic and community organizations), modes of movement, ecological factors, infrastructure, architecture, communications technologies, and linguistic systems. In other words, there are numerous factors both internal and external to a landscape that can initiate transformations or structural organizations. These can be abetted by human and non-human agents or can emerge from the system as a whole.

Bibliography

Adams, Thomas. *Garden City and Agriculture: How to Solve the Problem of Rural Depopulation*. Hitchin: Garden City Press Ltd., 1905.

———. *Recent Advances in Town Planning*. London: J. & A. Churchill, 1932.

Aldrich, H.E. *Organizations and Environments*. Englewood Cliffs: Prentice-Hall Inc., 1979.

Ashkenas, R., D. Ulrich, T. Jick, and S. Kerr. *The Boundaryless Organization: Breaking the Chains of Organizational Structure*. San Francisco: Jolley-Bass, 2002.

Ball, P. *Flow*. Oxford: Oxford University Press, 2009.

Bellamy, Edward. *Looking Backward, 2000-1887*. Harmondsworth: Penguin Books Ltd., 1982.

Benevolo, Leonardo. *History of Modern Architecture*, Volume 1. Cambridge, Mass.: MIT Press, 1971.

Berry, Wendell. *The Unsettling of America: Culture and Agriculture*. San Francisco: Sierra Club Books, 1986.

Braudel, Fernand. *Capitalism and Material Life 1400-1800*. London: Weidenfeld and Nicolson, 1973.

Brody, Hugh. *The Other Side of Eden: Hunters, Farmers and the Shaping of the World*. Vancouver: Douglas & McIntyre Ltd., 2000.

Buder, Stanley. *Visionaries and Planners: The Garden City Movement and the Modern Community*. New York: Oxford University Press, 1990.

Butor, Michel. *Frontiers*. Birmingham, Alabama: Summa Publications, Inc., 1989.

Castells, Manuel. *The Informational City: Information Technology, Economic Restructuring, and the Urban-Regional Process*. Oxford: Basil Blackwell, 1989).

Colebrook, Claire. *Gilles Deleuze*. London: Routledge, 2002.

Cooper, David E. *A Philosophy of Gardens*. Oxford, Clarendon Press, 2006.

Cooper, Thomas C. *Odd Lots: Seasonal Notes of a City Gardener*. New York: Henry Holt and Company, 1995.

Cranz, Galen. *The Politics of Park Design: A History of Urban Parks in America*. Cambridge, Mass.: MIT Press, 1982.

Creese, Walter L., ed. *The Legacy of Raymond Unwin: A Human Pattern for Planning*. Cambridge, Mass.: MIT Press, 1967.

———. "The Search for Environment". *The Garden City: Before and After*. Baltimore: Johns Hopkins University Press, 1992.

Deleuze, Gilles. *Foucault*. Trans. S. Hand. Minneapolis: University of Minnesota Press, 1988.

Deleuze, Gilles and Félix Guattari. *Kafka: Toward a Minor Literature*. Trans. D. Polan. Minneapolis: University of Minnesota Press, 1986.

———. *Anti-Oedipus: Capitalism and Schizophrenia*. Trans. R. Hurley, M. Seem and H.R. Lane. Minneapolis: University of Minnesota Press, 1983.

———. *A Thousand Plateaus: Capitalism and Schizophrenia*. Trans. B. Massumi. Minneapolis: University of Minnesota Press, 1987.

DeLanda, Manuel. *A New Philosophy of Society*. London: Continuum, 2006.

Fishman, Robert. *Urban Utopias in the Twentieth Century*. New York: Basic Books, Inc., 1977.

Forman, Richard T.T. and Michel Godron. *Landscape Ecology*. New York: John Wiley & Sons, 1986.

Forman, Richard T.T. *Land Mosaics: The Ecology of Landscapes and Regions*. Cambridge: Cambridge University Press, 1995.

George, Henry. *Progress and Poverty: An Inquiry into the Cause of Industrial Depressions and of Increase of Want with Increase of Wealth, The Remedy* (New York: The Modern Library, 1938).

Hall, Peter and Colin Ward. *Sociable Cities: The Legacy of Ebenezer Howard*. Chichester: John Wiley & Sons Ltd., 1998.

Hall, Peter. *Cities of Tomorrow: An Intellectual History of Urban Planning and Design in the Twentieth Century*. Oxford: Basil Blackwell, 1988.

Howard, Ebenezer. *Garden Cities of To-Morrow*. London: Faber and Faber Ltd., 1960.

———. *To-Morrow: A Peaceful Path to Real Reform*. London: Routledge, 2003.
Jacobs, Jane. *The Economy of Cities*. New York: Random House, 1969.
Jackson, Frank. *Sir Raymond Unwin: Architect, Planner and Visionary*. London: A Zwemmer Ltd., 1985.
Kropotkin, Peter. *Fields, Factories and Workshops, or, Industry Combined with Agriculture and Brain Work with Manual Work*. London: Thomas Nelson & Sons, 1913.
MacFadyen, Dugald. *Sir Ebenezer Howard and the Town Planning Movement*. Cambridge, Mass: MIT Press, 1970.
Maddren, M., ed. *Letchworth Recollections*. Wakefield: Egon Publishers Ltd., 1995.
Massingham, H.J. *This Plot of Earth: A Gardener's Chronicle*. London: Collins, 1944.
Masumoto, D.M. *Epitaph for a Peach: Four Seasons on My Family Farm*. New York: Harper Collins, 1996.
Meacham, S. *Regaining Paradise: Englishness and the Early Garden City Movement*. New Haven: Yale University Press, 1999.
McGrath, B., V. Marshall, M.L. Cadenasso, J. M. Grove, S.T.A. Pickett, R, Plunz, and J. Towers, eds. *Designing Patch Dynamics*. New York: GSAPP, Columbia University, 2007.
Miller, Mervyn. *Letchworth: The First Garden City*. Chichester: Phillimore & Co. Ltd., 1989.
———. *Raymond Unwin: Garden Cities and Town Planning*. Leicester: Leicester University Press, 1992.
Montmarquet, J.A. *The Idea of Agrarianism: Form Hunter-Gatherer to Agrarian Radical in Western Culture*. Moscow: University of Idaho Press, 1989.
Mumford, Lewis. *The Culture of Cities*. New York: Harcourt Brace Jovanovich, Inc., 1966.
Ohno, T. *Toyota Production System: Beyond Large-Scale Production*. Cambridge, Mass.: Productivity Press, 1988.
Orwin, C.S. *A History of English Farming*. London: Thomas Nelson and Sons Ltd., 1949.
Osborn, F.J. *Green-Belt Cities: The British Contribution*. London: Faber and Faber Ltd., 1946.
Parsons, K.C. and D. Schuyler. *From Garden City to Green City: The Legacy of Ebenezer Howard*. Baltimore: Johns Hopkins University Pres, 2002.
Pickett, S.T.A. and P.S. White, eds. *The Ecology of Natural Disturbance and Patch Dynamics*. Orlando: Academic Press, Inc., 1985.
Pollan, Michael. *Second Nature: A Gardener's Education*. New York: Grove Press, 1995.
Prescott, J.R.V. *The Geography of Frontiers and Boundaries*. Chicago: Aldine Publishing Co., 1965.
Purdom, C.B. *The Garden City: A Study in the Development of a Modern Town*. London: J.M. Dent & Sons Ltd., 1913.
———, ed. *Town Theory and Practice*. London: Benn Bros. Ltd., 1921.
———. *The Building of Satellite Towns: A Contribution to the Study of Town Development and Regional Planning*. London: J.M. Dent & Sons Ltd., 1949.
———. *Life Over Again*. London: J.M. Dent & Sons Ltd., 1951.
Simpson, M. *Thomas Adams and the Modern Planning Movement: Britain, Canada and United States, 1900-1940*. London: Mansell, 1985.
Spencer, Herbert. *Social Statics, or the Conditions Essential to Human Happiness Specified and the First of The Developed*. New York: Augustus M. Kelly Publishers, 1969.
Spirn, Anne Whiston. *The Granite Garden: Urban Nature and Human Design*. New York: Basic Books, 1984.
Stivale, Charles J., ed. *Gilles Deleuze: Key Concepts*. Montreal & Kingston; McGill-Queen's University Press, 2005.
Unwin, Raymond. *Town Planning in Practice: An Introduction to the Art of Designing Cities and Suburbs*. New York: Benjamin Blom, Inc., 1971.
Vogel, Steven. *Life in Moving Fluids: The Physical Biology of Flow*. Princeton: Princeton University Press, 1983.
Ward, S.V., ed. *The Garden City: Past, Present and Future*. London: E & FN Spon, 1992.
Williams, Raymond. *The Country and the City*. St. Albans: Paladin, 1975.

www.ingramcontent.com/pod-product-compliance
Lightning Source LLC
Chambersburg PA
CBHW021951290426
44108CB00012B/1032